Advances in Structural Integrity and Failure

Edited by Alireza Bahrami and Kenneth Imo-Imo Israel Eshiet

Published in London, United Kingdom

Advances in Structural Integrity and Failure
http://dx.doi.org/10.5772/intechopen.1000457
Edited by Alireza Bahrami and Kenneth Imo-Imo Israel Eshiet

Contributors
Abdul Rahim Sabouni, Sanin Dzidic, Aldin Mahmutovic, Alireza Khalifeh, Habib Danesh Manesh, Abolghasem Dehghan, Omid Karimzade Soureshjani, Ali Massumi, Siming Tian, Yang Xue, Zhenyan Tian, Shaoshuai Shi, Weidong Guo, Wei Wang, Ruijie Zhao, Estanislao Pujades, Miao Jing, Chunhui Lu, Anna Jurado, Yongxing Song

First published in London, United Kingdom, 2024 by IntechOpen
IntechOpen is the global imprint of INTECHOPEN LIMITED, registered in England and Wales, registration number: 11086078, 5 Princes Gate Court, London, SW7 2QJ, United Kingdom
Printed in Croatia

British Library Cataloguing-in-Publication Data
A catalogue record for this book is available from the British Library

Additional hard and PDF copies can be obtained from orders@intechopen.com

Advances in Structural Integrity and Failure
Edited by Alireza Bahrami and Kenneth Imo-Imo Israel Eshiet
p. cm.

This title is part of the Civil Engineering Book Series, Volume 5
Topic: Structural Engineering
Series Editor: Assed Haddad
Topic Editor: Wael A. Zatar

Print ISBN 978-1-83769-553-9
Online ISBN 978-1-83769-552-2
eBook (PDF) ISBN 978-1-83769-554-6
ISSN 3029-0287

IntechOpen Book Series

Civil Engineering

Volume 5

Aims and Scope of the Series

Civil engineering is a traditional field of engineering from which most other branches of engineering have evolved. It comprises traditional sub-areas like transportation, structures, construction, geotechnics, water resources, and building materials. It also encompasses sustainability, risk, environment, and other concepts at its core. Historically, developments in civil engineering included traditional aspects of architecture and urban planning as well as practical applications from the construction industry. Most recently, many elements evolved from other fields of knowledge and topics like simulation, optimization, and decision science have been researched and applied to increase and evolve concepts and applications in this field. Civil engineering has evolved in the last years due to the demands of society in terms of the quality of its products, modern applications, official requirements, and cost and schedule restrictions. This series addresses real-life problems and applications of civil engineering and presents recent, cutting-edge research as well as traditional knowledge along with real-world examples of developments in the field.

Meet the Series Editor

Professor Assed N. Haddad is a Civil Engineer with a degree from the Federal University of Rio de Janeiro (UFRJ) earned in 1986, as well as a Juris Doctor degree from the Fluminense University Center earned in 1993, and a Master's degree in Civil Engineering from the Fluminense Federal University (UFF) obtained in 1992. He completed his Ph.D. in Production Engineering from COPPE / Federal University of Rio de Janeiro in 1996. Professor Haddad's academic pursuits have taken him to postdoctoral stays at the University of Florida, USA in 2006; at the Universitat Politècnica de Catalunya, Spain in 2010; and at the University of New South Wales Sydney, Australia in 2019. Currently, he serves as a Full Professor at the Federal University of Rio de Janeiro. He has held visiting professorships at various institutions including the University of Florida, Universitat Politècnica de Catalunya, Universitat Rovira i Virgili, and Western Sydney University. His research expertise encompasses Civil, Environmental, and Production Engineering, with a primary focus on the following topics: Construction Engineering and Management, Risk Management, and Life Cycle Assessment. He has been the recipient of research grants from the State of Rio de Janeiro, Brazil: CNE FAPERJ from 2019 to 2022 and from 2023 to 2025. Additionally, his research grants obtained from the Brazilian Government CNP since 2012 last to this date. Professor Haddad has been involved in several academic endeavors, being the Guest Editor of the International Journal of Construction Management; MDPI's Sustainability, Energies, and Infrastructures; Associate Editor at Frontiers in Built Environment / Sustainable Design and Construction; Guest Editor at Frontiers in Built Environment / Construction Management; and Academic Editor of the Journal of Engineering, Civil Engineering Section of Hindawi. He is currently a Professor of the Environmental Engineering Program at UFRJ and the Civil Engineering Program at UFF.

Meet the Volume Editors

Dr. Alireza Bahrami is an Associate Professor of Civil and Structural Engineering and Head of Subject at the University of Gävle, Sweden. He has many years of experience in civil and structural engineering including academic experiences at several universities and practical (industrial) experiences at various consulting and construction companies. He has taught and supervised many different specialized courses for Ph.D., master's, and bachelor's students at universities. He has carried out numerous scientific research works and published book chapters and articles in internationally reputable journals and conferences. Dr. Bahrami has also practically worked on various projects for different companies.

Dr. Kenneth Imo-Imo Israel Eshiet is a senior lecturer in the Department of Civil Engineering, University of Wolverhampton, United Kingdom. In addition, he is the subject lead of the geotechnical engineering subject area and the program leader of the BEng Civil Engineering program. He has extensive experience working as an engineer and academic at several academic institutions and in industry. Amongst other skills, his professional work has provided him with broad exposure to and expertise in the modeling of natural and engineered systems and processes. He is a chartered engineer, a corporate member of the Institution of Civil Engineers (ICE), and a Fellow of Advance Higher Education (Advance HE), United Kingdom. He is also an active member of several other professional bodies. He holds a doctorate degree in Civil Engineering from the University of Leeds in tandem with other postgraduate qualifications in management and education. His current interest involves the design of engineering systems/structures, and the development and application of numerical/analytical tools to civil, structural, fluid dynamics, and geotechnical engineering problems.

Contents

Preface

This book discusses structural integrity and failure and underground construction. In the realm of civil and structural engineering, the pursuit of structural integrity stands as a paramount goal. Structures, whether towering skyscrapers or intricate bridges are the physical manifestations of human ingenuity and technological prowess. However, the inherent complexity of these creations makes them susceptible to the forces of nature, the passage of time, and the ever-present potential for failure.

At its core, structural integrity refers to the ability of a system or component to withstand its intended load without experiencing deformation or failure. This concept goes beyond merely preventing collapse; it encompasses the ability of a structure to maintain its form and function over time. Achieving structural integrity involves a delicate balance of materials, design, and construction techniques.

Materials play a pivotal role in determining a structure's ability to resist external forces. Engineers meticulously select materials based on their strength, elasticity, durability, and other essential properties. The design phase is equally critical, requiring a deep understanding of physics, mathematics, and the specific demands placed on the structure. The construction process also demands precision and adherence to stringent standards to ensure that the designed integrity is faithfully realized.

Structures exist within a dynamic environment where external forces constantly act upon them. Forces such as wind, earthquakes, and gravity exert their effects, testing the resilience of the built environment. Nature's unpredictable elements challenge engineers to anticipate and mitigate potential weaknesses. Moreover, the relentless march of time introduces the concept of fatigue, corrosion, and wear, necessitating ongoing maintenance and monitoring.

Failure, though often considered an undesirable outcome, serves as a powerful teacher in the pursuit of structural integrity. Historical examples of structural failures, from the collapse of bridges to the buckling of buildings, offer valuable lessons. Each failure contributes to a collective body of knowledge, prompting the engineering community to refine standards, improve materials, and enhance construction methodologies.

In the contemporary era, technology plays a pivotal role in ensuring and assessing structural integrity. Advanced simulation tools, computer-aided design, and real-time monitoring systems empower engineers to model and predict the behavior of structures under diverse conditions. These tools aid in optimizing designs, identifying potential weaknesses, and implementing preventive measures.

As we delve into the exploration of structural integrity and failure, we embark on a journey that traverses the intersection of science, art, and practicality. The stories of triumph and tragedy in the world of engineering underscore the importance of continuous improvement and a deep commitment to the safety and sustainability of the built environment.

As our urban landscapes evolve and population densities increase, the demand for efficient and sustainable use of space has never been more critical. Underground construction emerges as a compelling solution, offering a realm of possibilities that extend beyond conventional building practices. Tunnels and subways represent just a fraction of the diverse applications that harness the potential beneath our feet.

The challenges of underground construction are multifaceted, encompassing geological considerations, engineering precision, and innovative technologies. Excavating through diverse geological formations requires a delicate balance of knowledge and adaptability. Engineers must navigate through solid rock, soft soil, or complex strata, often employing cutting-edge techniques to overcome geological hurdles.

Beyond the technical aspects, underground construction also addresses environmental concerns and sustainability. By moving critical infrastructure and transportation networks below ground, we not only optimize land use but also minimize the impact on the surface ecosystem. The subterranean space becomes a sanctuary for preserving green spaces, historical landmarks, and urban aesthetics.

This book consists of seven chapters in two sections. Section 1, "Structural Integrity and Failure", includes Chapter 1: "Advances in Reinforced Concrete Integrity and Failure", Chapter 2: "Proposing a Design Model for Determining Flexural Bearing Capacity of RC Beams Reinforced by Steel with Reduced Modulus of Elasticity", Chapter 3: "Fracture Behavior of Structural Steels under Earthquake Dynamic Loadings", and Chapter 4: "Seismic Assessment of RC Shear Wall Structure under Real Seismic Sequences Using Equivalent Shell-Wire Model".

Section 2, "Underground Construction" includes Chapter 5: "Design Method and Construction Technology in Tunnel Engineering under Complex Geological Conditions", Chapter 6: "Underground Excavations below the Water Table by the Cut-and-Cover Method", and Chapter 7: "On-Line Monitoring and Intelligent Diagnosis Technology of Rail Transit Ventilation System".

Alireza Bahrami
Faculty of Engineering and Sustainable Development,
Department of Building Engineering,
Energy Systems and Sustainability Science,
University of Gävle,
Gävle, Sweden

Kenneth Imo-Imo Israel Eshiet
Department of Civil Engineering,
University of Wolverhampton,
Wolverhampton, England

Section 1

Structural Integrity and Failure

Chapter 1

Advances in Reinforced Concrete Integrity and Failure

Abdul Rahim Sabouni

Abstract

The chapter provides an overview of the advances made in the field of structural integrity and failure with a specific focus on reinforced concrete structures. It begins with a brief history of reinforced concrete and covers its structural properties and characteristics. It then delves into the mechanics of reinforced concrete structures, including the various forces that act on them, and the design and construction of these structures. It delves into the basic mechanics, stressing the concrete's performance under loading and its inherent material properties. The focus then shifts to the design principles applied to reinforced concrete structures, and the consideration of critical structural elements like beams, slabs, columns, and foundations. Various advances in reinforced concrete technology, including High-Performance Concrete, Fiber-Reinforced Concrete, Self-Compacting Concrete, and the use of nanomaterials, are explored. The chapter provides insights into methods for the analysis and assessment of reinforced concrete structures, discussing non-destructive testing methods, structural health monitoring, and finite element analysis. It examines the causes of failure, including material quality, overloading, design flaws, environmental factors, and construction errors. Several case studies of notable building failures are highlighted, emphasizing lessons learned and the importance of safe construction practices. The chapter concludes by looking at future directions in reinforced concrete, encompassing advanced materials, digital technology, sustainable construction practices, and resilience-based design.

Keywords: advancement in reinforced concrete, reinforced concrete construction, structural concrete integrity, reinforced concrete failures, progressive collapse, reinforced concrete materials

1. Introduction

1.1 Brief overview of reinforced concrete

Reinforced concrete is a composite material where the concrete, possessing high compressive strength but relatively weak in tension, is combined with materials that excel in withstanding tensile stress. The primary objective of such reinforcement is to compensate for concrete's inherent weakness in tension. When tensile stresses exceed the tensile strength of the concrete, cracks occur. If these stresses persist, the cracks can propagate, potentially leading to the failure of the structure.

The material of choice for reinforcement has been and remains to be steel. The high tensile strength of steel and its exceptional bond with concrete are the main reasons for its widespread use. When encased in concrete, the steel reinforcement acts to counteract tensile stresses, while the concrete resists compressive stresses. This synergetic effect of two materials working together to resist different kinds of stresses is one of the key reasons behind the widespread use of reinforced concrete in the construction industry [1].

Over the years, reinforced concrete has proven to be a versatile, durable, and cost-effective construction material. It is utilized in a wide array of structures including buildings of all types and sizes, bridges, dams, tunnels, roads, and many more. Concrete is the most consumed construction material, and twice as much concrete is used in construction as all other building materials combined [2].

1.2 Importance of reinforced concrete in structural integrity

Structural integrity is a key concept in the field of civil engineering, referring to the ability of a structure to withstand its intended load without experiencing distress or failure. Maintaining structural integrity is a primary concern in the design, construction, and maintenance phases of a structure.

The nature of reinforced concrete—being a composite material that combines the compressive strength of concrete with the tensile strength of the reinforcement—plays a critical role in ensuring structural integrity. The result is a material that exhibits high robustness and resilience under a variety of load conditions, environmental influences, and potential man-made and natural disasters [3].

Structural integrity is particularly critical for structures exposed to dynamic loads and changing environmental conditions. This includes structures like bridges, which must resist the dynamic loads of passing vehicles; skyscrapers, which must withstand wind loads; and coastal structures, which must resist the destructive forces of waves and salt corrosion. The combination of concrete and steel reinforcement creates a material that can effectively resist these and other forces, thereby maintaining the structural integrity of these constructions.

Moreover, the importance of structural integrity extends beyond the avoidance of catastrophic structural failures. It also influences other factors such as the longevity of structures, maintenance costs, and even esthetic appeal. For instance, cracks in concrete not only compromise its structural integrity, but can also accelerate the ingress of harmful substances like water, chloride, and carbon dioxide, leading to further deterioration and higher maintenance costs. Similarly, visible cracks and deflections can detract from the esthetic appeal of a structure, which can be especially problematic for architectural or heritage structures [4, 5].

In this context, the role of reinforced concrete in maintaining structural integrity becomes even more prominent. Advancements in reinforced materials, design methods, and construction techniques are reflected with parallel advancements in the design, assessments, and quality assurance of structural integrity [6–8].

2. Historical perspective

2.1 Evolution of reinforced concrete

Reinforced concrete's rich history can be traced back to the mid-19th century. The first widespread use of concrete reinforcement is often credited to the French industrialist François Coignet in 1854. Coignet recognized the potential of embedding iron bars into concrete to create a composite material that capitalized on the strengths of both components [9].

Joseph Monier, a French gardener, took this idea a step further. Looking for a way to create more durable flowerpots, Monier discovered that reinforcing the concrete used to form the pots enhanced their strength and durability. Recognizing the broader implications of his innovation, Monier patented the use of reinforced concrete in 1867 [10].

The evolution of reinforced concrete continued throughout the late 19th and early 20th century, driven by the pioneering work of many visionaries. François Hennebique, a self-taught French engineer and builder, obtained a patent for his groundbreaking system of reinforced-concrete construction in 1892. This system uniquely merged different components of construction, like the column and beam, into a single, unified piece. The Hennebique system was instrumental in paving the way for modern methods of reinforced-concrete construction, being one of its earliest iterations. He developed the technique of systematic reinforcement, optimizing the placement and orientation of the reinforcement within the concrete [11].

Another notable figure, Swiss engineer Robert Maillart, pushed the esthetic and structural possibilities of reinforced concrete. Maillart is particularly known for his revolutionary bridge designs, which combined structural efficiency with artistic elegance [12, 13].

2.2 Key milestones and innovations

The 20th century marked numerous innovations that revolutionized the use and applications of reinforced concrete. A key breakthrough was the development of pre-stressed concrete by French engineer Eugène Freyssinet in the 1920s. By pre-tensioning the steel reinforcement before pouring the concrete, Freyssinet found that it was possible to create long-span concrete structures with reduced depth, opening new possibilities for the design of bridges, roofs, and other large-span structures [14, 15]. Pre-stressed concrete continued its development to play an important role in various sectors of modern construction [16].

The latter part of the 20th century saw the introduction of advanced admixtures that further improved the properties of concrete. By modifying the composition of concrete, these admixtures made it more durable, workable, and adaptable to various applications [17, 18].

Another significant development was the introduction of fiber-reinforced polymers (FRP) as an alternative to traditional steel reinforcement. Unlike steel, FRP does not corrode, making it particularly suited for use in harsh environments such as marine structures or chemical plants [19, 20].

The 21st century continues to witness numerous advancements in the field of reinforced concrete. Innovations in high-performance concrete, ultra-high-performance concrete [21, 22], self-healing concrete [23], and the use of nanomaterials are pushing the boundaries of what is achievable with this versatile material [24].

As we move forward, the ongoing advancements in the field of reinforced concrete promise to continue transforming our built environment, enhancing its safety, efficiency, and sustainability. The following sections will delve into the fundamental principles, material properties, design considerations, case studies, and recent advancements in the field of reinforced concrete, thereby providing a comprehensive understanding of its critical role in ensuring structural integrity and preventing failure.

3. Fundamental principles of reinforced concrete

3.1 Basic mechanics

The study of reinforced concrete requires a clear understanding of its basic mechanics. This often involves delving into the stress–strain relationships of the materials involved, typically concrete and steel, which display starkly different mechanical behaviors.

3.1.1 Stress: strain characteristics

Stress–strain curves, or relationships, depict how a particular material responds to applied stress. Stress denotes the force exerted per unit area, while strain represents the resulting deformation. Understanding these curves is pivotal to manipulating and predicting a material's performance under various loading conditions [1].

Concrete, as a material, displays exceptional strength when subjected to compressive loads, but it lacks tensile strength. Consequently, it exhibits a nonlinear stress–strain relationship (**Figure 1**). Under compressive loading, concrete experiences peak stress at its maximum compressive strength. Any additional stress results in a decrease, signaling material failure.

The opposite is true for steel, a material commonly used for reinforcement due to its strong tensile and compressive strengths [2].

The stress–strain relationship for steel is linear up to the yield strength, the point at which steel undergoes plastic deformation. Beyond the yield strength, steel can experience significant strain without an accompanying increase in stress. This key difference underlines the complementarity of concrete and steel in reinforced concrete, leading to its wide usage in construction (see **Figure 2**).

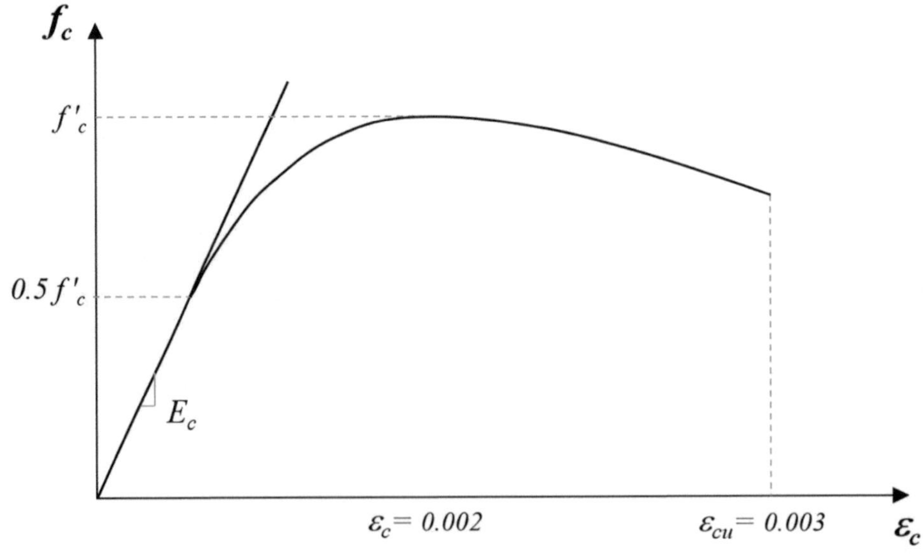

Figure 1.
Typical stress–strain curve for concrete.

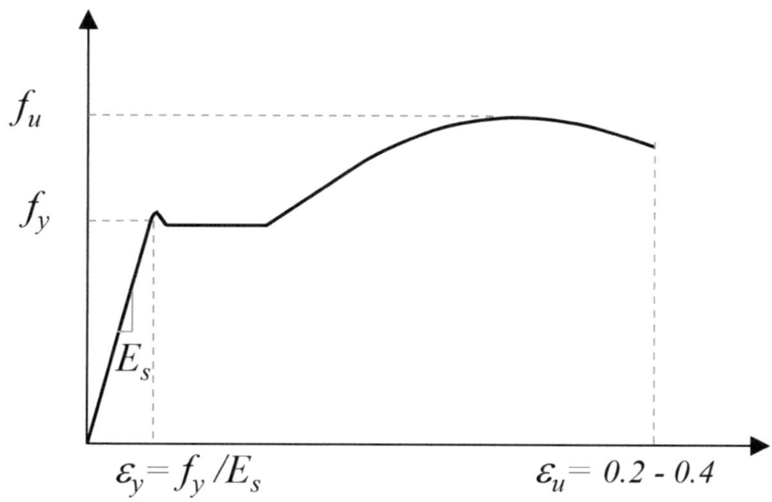

Figure 2.
Typical stress–strain curve for steel reinforcement.

3.1.2 Flexure, shear, and torsion

Reinforced concrete structures are subjected to a variety of stresses, including flexural, shear, and torsional stresses. Flexural stress is generated when a member is subjected to bending, inducing tension on one side and compression on the other. Shear stress, in contrast, is caused by forces acting parallel to the cross-sectional area, leading to a sliding failure of the material. Torsional stress occurs due to twisting forces [2].

The design of reinforced concrete takes into account these diverse stress types, necessitating the calculation and precise arrangement of reinforcement to counteract these forces effectively.

3.2 Behavior under loading

Understanding the behavior of reinforced concrete under loading is crucial to both its design and analysis. This behavior can be categorized into three main domains: cracking, deflection, and load capacity.

3.2.1 Cracking

Cracking is a fundamental characteristic of concrete, resulting from its brittleness and low tensile strength. Initial cracks often form due to shrinkage as the concrete cures, with further cracks appearing due to applied loads. The reinforcement in reinforced concrete structures serves to tightly bind these cracks, preventing sudden and brittle failure.

The thorough study of cracking patterns and their causative factors can inform construction practices, leading to better control of cracking and the development of more robust and durable structures. This requires a comprehensive understanding of the material properties of both concrete and the reinforcement, along with the environmental and loading conditions the structure is expected to encounter [2]. Correct cracking assessment may require evaluation and consideration of multiple causes and has led to the utilization of Artificial Intelligence and development of expert systems for the crack assessment and recommendations of proper structural integrity solutions for distressed reinforced concrete structures [25, 26].

3.2.2 Deflection

Deflection refers to the displacement of a structural element under a load. It is a critical aspect to monitor, as excessive deflection can lead to a host of serviceability issues. These can range from esthetic concerns, such as cracking in finishes, to functional issues, like ponding on roofs and unpleasant vibrations.

The deflection of a reinforced concrete member depends on its geometry, the properties of the materials, and the nature of the load applied. Careful consideration of these factors during the design process can help to manage deflection and avoid potential problems down the line [1].

3.2.3 Load capacity

Load capacity is the maximum load that a reinforced concrete member can sustain without experiencing failure. It is determined by considering the section's geometry, the properties of the materials, and the nature of the load applied. In the design of reinforced concrete structures, the load capacity must be calculated for various limit states to ensure that it is never exceeded during the structure's intended service life [2].

Accurately calculating load capacity requires a thorough understanding of reinforced concrete behavior under different loading and environmental conditions. Additionally, understanding failure modes and recognizing the signs of imminent failure can greatly aid in maintaining the structural integrity of a reinforced concrete structure.

The fundamental principles of reinforced concrete, as outlined in this section, form the basis for understanding its material properties, design considerations, and innovative advancements in the field. The following sections will delve into these topics, providing a comprehensive understanding of reinforced concrete's role in ensuring structural integrity and preventing failure.

4. Material properties

4.1 Concrete

4.1.1 Composition and strength

Concrete is a composite material that is composed of coarse aggregate (usually gravel or crushed stone), fine aggregate (often sand), and a binder (cement), all of which are mixed with water. The relative proportions of these components play a significant role in determining the final properties of the concrete.

The strength of concrete is primarily a function of the water-to-cement (w/c) ratio, the type of cement used, and the aggregates used. A lower w/c ratio tends to result in a concrete mixture that possesses higher strength but lower workability. Conversely, a high w/c ratio increases the workability of the concrete but at the expense of reducing its strength.

Several other factors, such as the curing conditions and the presence of admixtures, can also influence the strength of concrete. These factors can be controlled and adjusted to produce concrete with specific desired properties.

4.1.2 Durability issues

Durability is a measure of the ability of concrete to resist various forms of deterioration. This could include damage from freeze–thaw cycles, chemical attacks, and abrasion. The durability of concrete is influenced by several factors, including the w/c ratio, the type and quantity of cement used, the type and size of aggregates used, and the use of certain admixtures.

One of the most significant durability issues for reinforced concrete is the corrosion of reinforcing steel. In highly alkaline conditions, concrete can passivate the steel, protecting it from corrosion. However, if the alkalinity is reduced due to processes such as carbonation, or if chloride ions penetrate the concrete and reach the steel, corrosion can occur.

4.2 Reinforcement

4.2.1 Steel reinforcing bars

Steel reinforcing bars, commonly referred to as rebar, are the most frequently used type of reinforcement in concrete. Rebar is available in several grades, with yield strengths typically ranging from 40,000 psi (275 MPa) to 100,000 psi (690 MPa) [2]. To enhance bonding with the concrete, the surface of the rebar is often deformed.

4.2.2 Fiber-reinforced polymers (FRP)

Fiber-Reinforced Polymers (FRP) have emerged as an alternative to steel rebar, especially in environments where corrosion is a significant concern. FRP bars are composed of continuous fibers (such as glass, carbon, or aramid) embedded in a polymer matrix. They are non-corrosive, lightweight, and have a high tensile strength. However, they are also more brittle than steel and have a lower modulus of elasticity [27]. Moreover, researchers have used basalt FRP bars and investigated their bond characteristics with concrete [28].

In the upcoming sections, we will delve deeper into the design considerations for reinforced concrete structures and explore the latest advancements in the field.

5. Design of reinforced concrete structures

5.1 Design approaches

5.1.1 Limit state design

Limit state design is a structured approach used in structural engineering to ensure the safety and functionality of structures under all conceivable conditions. The concept of limit state design revolves around two primary types of limit states: ultimate limit states (ULS) and serviceability limit states (SLS) [29, 30].

The ULS pertains to conditions that would result in the failure or collapse of the structure, such as yielding, buckling, or fracture. These situations usually involve severe loading conditions or structural deterioration that exceeds the structure's load-bearing capacity. The intent of designing for ULS is to ensure the structure can resist severe or extreme loads without collapse, thereby providing a high level of safety.

The SLS, conversely, relates to conditions that do not cause a structural collapse but can negatively affect the structure's appearance, durability, and usability. This can include conditions like excessive deflection, vibration, and cracking. While these situations may not immediately threaten the structure's safety, they can diminish the structure's serviceability and lifespan.

The overarching aim of limit state design is to ensure that the structure will not reach any of these limit states during its expected service life. This is achieved by applying appropriate safety factors to the loads and material strengths, considering potential variations and uncertainties.

5.1.2 Performance-based design

The performance-based design represents a more contemporary and sophisticated approach to structural design. Instead of strictly adhering to prescriptive code requirements, performance-based design involves specifying the desired performance levels of a structure under different load scenarios [31].

This approach requires a more nuanced understanding of the structure's behavior and often involves advanced analysis techniques. Despite its complexity, performance-based design can yield more efficient and optimized structures by allowing more flexibility in the design process and better alignment of the design with the intended function and performance of the structure [32].

5.2 Basic structural elements

5.2.1 Beams and slabs

Beams and slabs are integral horizontal structural elements, primarily designed to resist bending. The role of the reinforcement in beams and slabs is to strategically counteract the tensile stresses induced by bending. Meanwhile, the concrete component resists the compressive stresses.

5.2.2 Columns

Columns, being vertical structural elements, are primarily designed to resist axial compression. The design of columns involves ensuring that the column can safely carry the applied load without buckling or crushing. Additionally, columns often need to resist bending due to eccentric loading or lateral forces, which requires additional reinforcement.

5.2.3 Foundations

The role of foundations is to transfer the loads from the structure to the ground effectively. The design of foundations involves assessing the ground conditions to ensure it can safely support the imposed loads without causing excessive settlement or failure. Reinforced concrete is commonly used for various types of foundations, including spread footings, strip footings, and pile caps.

In the subsequent sections, we will explore the recent advancements in the field of reinforced concrete design and discuss the various methods employed for assessing and improving the structural integrity and potential failure of reinforced concrete structures.

6. Advances in reinforced concrete

6.1 High-performance concrete and ultra-high-performance concrete

High-Performance Concrete (HPC) is an advanced type of concrete that has been meticulously engineered to possess superior properties such as exceptional strength, extended durability, or improved workability. The attributes of HPC are typically achieved through a low water-to-cement ratio in conjunction with the inclusion of admixtures and supplementary cementitious materials (SCMs) like fly ash, silica fume, or slag [33, 34].

HPC often exceeds a compressive strength of 6000 psi (41 MPa), significantly higher than the average 3000 psi (20 MPa) for regular concrete. This high strength allows for the design and construction of slenderer and efficient structures. Moreover, HPC's enhanced durability can contribute to a longer service life and potentially lead to reduced maintenance costs [34].

Ultra-High-Performance Concrete (UHPC) is characterized by its significantly superior strength and durability compared to traditional concrete, UHPC typically surpasses a compressive strength of 21,000 psi (145 MPa). This level of strength makes it suitable for high-stress applications and for constructing structures that are highly resistant to environmental and mechanical stresses [35].

6.2 Fiber-reinforced concrete

Fiber-Reinforced Concrete (FRC) is a unique form of concrete where small fibers are incorporated into the concrete mix to enhance its properties. These fibers can be derived from a variety of materials, including steel, glass, synthetic materials, or natural substances. The fibers serve to enhance the concrete's tensile strength, ductility, and resistance to cracking and shrinkage. Due to these improved characteristics, FRC has found applications in several areas, including overlays, precast elements, and structures exposed to dynamic or impact loads [35].

6.3 Self-compacting concrete

Self-Compacting Concrete (SCC) is a distinctive type of concrete that possesses the ability to flow and consolidate under its own weight, thereby eliminating the need for mechanical vibration. This property makes SCC especially beneficial for complex forms or densely reinforced structures. The key to achieving self-compacting properties lies in a high content of fine materials and the use of high-range water reducers, commonly known as superplasticizers [36].

6.4 Use of nanomaterials

The potential of nanomaterials in revolutionizing the field of concrete technology has been an exciting recent development. Research has indicated that nanomaterials like nano-silica, nano-titanium oxide, and carbon nanotubes can significantly enhance the properties of concrete. These improvements can span across concrete's strength, durability, and even grant it self-cleaning properties [37]. However, the application of nanomaterials in concrete is still in its nascent stages. Further research is required to address challenges related to cost, optimal dispersion, and health and safety considerations [38].

6.5 Engineered cementitious composites

Engineered Cementitious Composites (ECC), also known as "bendable concrete," ECCs are specifically engineered to exhibit strain-hardening and multiple cracking behaviors under tensile stress, unlike traditional concrete which is brittle and prone to cracking. These composites incorporate a high volume of fibers, generally polymer-based, to enhance the ductility and durability of the material. This advanced composite material can significantly reduce maintenance costs and extend the service life of structures, proving to be a promising solution for sustainable infrastructure development [39].

6.6 Eco-concrete

As sustainability becomes a crucial aspect of construction, Eco-Concrete has emerged as a significant advancement. Eco-Concrete refers to any concrete that utilizes recycled aggregate, recycled or waste materials as part or all of its constituents, reducing the consumption of natural resources and CO2 emissions associated with cement production. Examples include the use of fly ash slag [40–42], recycled concrete aggregates [43, 44], reusing waste shredded tires [45], and waste glass. In addition to being environmentally friendly, research has shown that some forms of Eco-Concrete can also match or even exceed the performance of traditional concrete in terms of strength and durability [46, 47].

6.7 3D printed concrete

3D Printed Concrete has emerged as a radical advancement in the field of reinforced concrete, bringing about a revolution in the construction industry. This technology utilizes robotic arms and extrusion techniques to deposit layers of specially formulated concrete mix, enabling the creation of complex geometric forms that are difficult or even impossible to achieve with traditional construction methods. 3D-printed concrete can offer significant benefits including cost reduction, speed of construction, design flexibility, and the reduction of construction waste [48].

6.8 Bacterial concrete

Bacterial Concrete, also known as Bio-Concrete, is an innovative solution to the issue of concrete durability. Certain types of bacteria, when introduced into the concrete mix, can induce calcium carbonate precipitation, which heals the micro-cracks that form in concrete over time. This self-healing property can significantly improve the durability and lifespan of concrete structures, reduce maintenance costs, and contribute to sustainability in the construction industry. Challenges such as ensuring long-term bacterial survival and minimizing cost remain to be addressed [23].

DOI: http://dx.doi.org/10.5772/intechopen.1002247

6.9 Structural health monitoring in concrete structures

Structural Health Monitoring (SHM) technologies are increasingly being integrated into reinforced concrete structures to assess their integrity and predict their performance over time. These technologies involve the use of sensors and data analysis algorithms to detect anomalies and deterioration in structures. With advancements in the Internet of Things (IoT) and machine learning, SHM has the potential to revolutionize the maintenance and inspection process of concrete structures, leading to significant savings in terms of time and costs, as well as enhancing overall safety [38].

6.10 Smart concrete

Smart Concrete is a type of concrete embedded with sensors capable of sensing changes in the material or surrounding conditions. These sensors can detect stresses, strains, temperature changes, moisture content, and even chemical changes, which can be crucial for real-time monitoring and early detection of potential failures. This advancement could pave the way for the development of intelligent infrastructure with greatly improved safety, durability, and maintenance efficiency [49].

7. Analysis and assessment of reinforced concrete structures

7.1 Non-destructive testing methods

Non-Destructive Testing (NDT) methods are crucial for assessing the health and integrity of structures without causing any damage to them. These techniques have evolved significantly over time and have been especially useful for inspecting concrete structures. NDT methods provide invaluable data regarding the condition of the structure, thereby helping identify potential issues before they escalate into serious problems.

Several NDT methods are in regular use, each having unique applications and capabilities. These include visual inspection, the use of a rebound hammer, ultrasonic pulse velocity (UPV), ground penetrating radar (GPR), and core sampling, among others.

7.1.1 Visual inspection

It is the most straightforward method of NDT, relying on the examiner's expertise to detect anomalies such as cracking, spalling, or signs of corrosion. This method is cost-effective and quick but can only detect superficial defects.

7.1.2 Rebound hammers

In a rebound hammer, a mass is loaded over a spring and can be moved axially along a tubular housing which imparts a defined amount of energy (**Figure 3**). As the plunger is pressed against the surface to be examined, the spring is extended from a fixed position. The mass rebounds, upon release, from the plunger which remains in contact with the concrete surface, the distance by which the mass rebounds is measured and presented as a percentage of the initial extension (while pressing

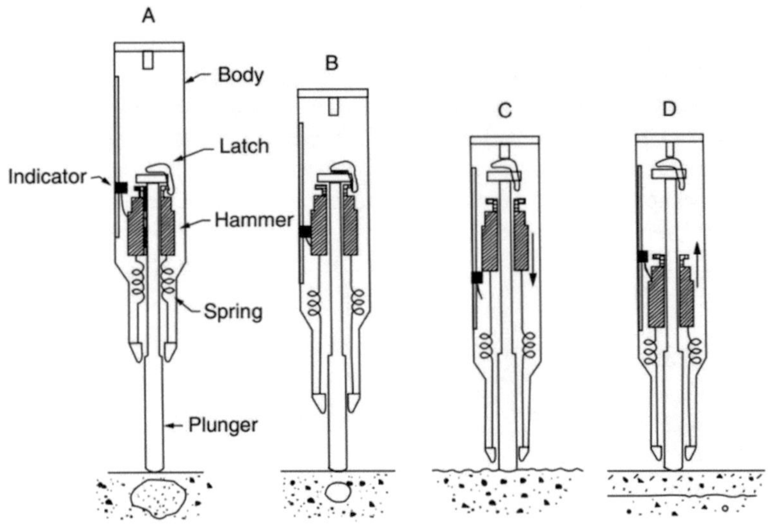

Figure 3.
Rebound hammer for non-destructive testing of compressive strength [50].

the hammer) of the spring. This percentage is known as rebound number. They are widely used due to their simplicity and portability. They measure the surface hardness of concrete, which can be correlated to its compressive strength. However, they only provide a qualitative assessment and can be influenced by surface condition and moisture content (**Figure 3**) [50].

7.1.3 Ultrasonic pulse velocity (UPV) tests

They involve sending an ultrasonic wave through a concrete element and measuring the time it takes to travel. This information can help determine the quality of the concrete and detect internal flaws. While effective, this method requires access to both sides of the concrete element (**Figure 4**) [51].

7.1.4 Ground penetrating radar (GPR) tests

They involve emitting electromagnetic waves into the concrete and capturing the reflected signals. This provides details about the location and depth of reinforcement bars, post-tension cables, and the presence of voids or delamination. Although GPR is a versatile tool, it requires skilled interpretation and can sometimes be hampered by interference from metallic objects.

This non-destructive testing (NDT) technique is highly efficient in the detection of metallic objects embedded in concrete or other non-conductive materials because of the high reflectivity of metals and the contrast between the electromagnetic properties of metals and concrete.

The application of GPR to the detection of corrosion and damage as a consequence of corrosion has demonstrated its effectiveness in the detection of corrosion by measuring qualitative changes in the amplitude, even though variation in depth could affect highly the results [52, 53].

DOI: http://dx.doi.org/10.5772/intechopen.1002247

Figure 4.
Ultrasonic pulse velocity tests [51].

7.1.5 Core sampling

It involves extracting small cylindrical samples from the structure for laboratory testing. Although it is the most reliable method for determining concrete's properties, it causes minor damage to the structure. Therefore, it is typically used when other NDT methods indicate potential issues that need further investigation (**Figure 5**).

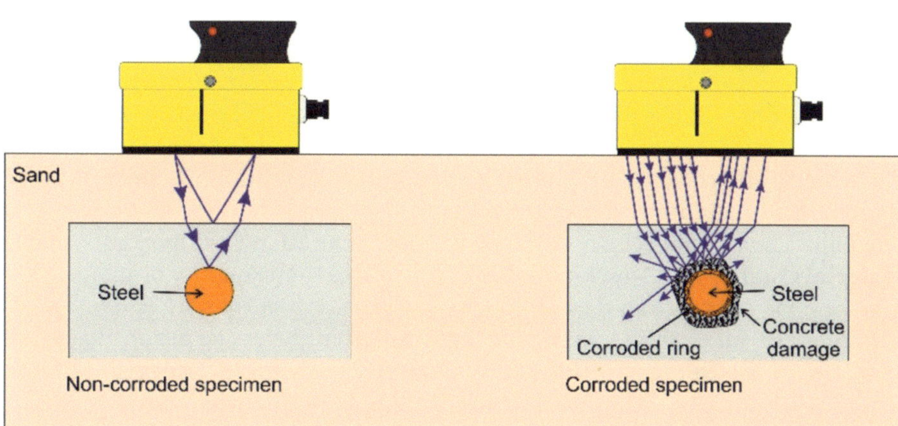

Figure 5.
Ground penetrating radar for non-destructive testing of compressive strength [53].

In practice, a combination of different NDT methods is often employed to gain a more comprehensive understanding of a structure's health. The choice of methods largely depends on the specific objectives of the inspection and the complexities of the structure itself. By providing vital data about factors like compressive strength, depth of cover, location of reinforcement, and the presence of defects or signs of deterioration, NDT methods play a crucial role in ensuring the longevity and safety of our concrete structures.

7.2 Structural health monitoring

Structural Health Monitoring (SHM) is a proactive approach that uses sensors to consistently monitor a structure's condition and performance over time. These sensors can measure a wide range of parameters, including strain, temperature, vibration, and corrosion rate, among others. The data gathered can be thoroughly analyzed to detect any changes or irregularities that may suggest an issue [38].

SHM has the advantage of providing early warnings of potential issues, thereby enabling timely maintenance or repair. As a result, the service life of the structure can be increased, and lifecycle costs can be reduced [54, 55].

7.3 Finite element analysis

Finite Element Analysis (FEA) is a powerful computer-based method used to predict how a structure will respond to various loads and conditions. FEA involves the division of the structure into small elements and then solving the governing equations for each element. The results are then combined to obtain the overall response of the structure [56, 57].

In the context of reinforced concrete structures, FEA can be used to analyze the complex interactions between the concrete and the reinforcement. This includes the nonlinear behavior of the materials, the bond between the concrete and the reinforcement, the cracking of the concrete, and the yielding of the reinforcement. By accurately modeling these phenomena, FEA can provide valuable insights into the behavior and performance of reinforced concrete structures under various load scenarios, including both service loads and extreme events.

8. Structural loss of integrity and failure of reinforced concrete structures

The main causes of structural integrity and failure in reinforced concrete structures can be grouped into various categories.

The following is what is believed to be the list of the 10 most important categories:

Material Quality and Workmanship: This relates to the quality of materials used in the construction and the skill with which those materials are employed. Poor quality concrete, for example, may have low strength or be prone to shrinkage, which can result in cracks. Similarly, low-quality or poorly placed reinforcement can lead to inadequate resistance to the applied forces, leading to premature failure.

Load and Overloading: This includes factors like dead load (weight of the structure itself), live load (weight of objects and occupants), snow load, wind load, and earthquake load. Overloading beyond the design load can lead to structural failure.

Design Flaws: If the structure has been designed without properly accounting for all loads or has been designed without enough redundancy, it can be susceptible to failure. This includes neglecting to design for shear forces or bending moments, misjudging the weight of materials or the potential live loads, or neglecting potential environmental effects.

Environmental Factors: This includes the impact of weather conditions, temperature changes, moisture (leading to corrosion of reinforcements or freeze–thaw cycles), chemical attacks (such as from chlorides, sulfates, etc.), and seismic activity.

Fatigue and Wear: This occurs due to repetitive loading over time, which can weaken the structure. Even if each individual load is within the design limit, the repetitive nature of the load can lead to fatigue failure.

Corrosion of Reinforcing Steel: This is one of the most common causes of deterioration in reinforced concrete structures. It can be caused by exposure to elements like water and oxygen, leading to the formation of rust. Corrosion expands the volume of the steel, which can cause concrete to crack and spall.

Construction Errors: These are mistakes or inaccuracies made during the construction phase, such as incorrect placement of reinforcement, improper mixing or pouring of concrete, or inadequate curing, which can lead to premature failure.

Deflection and Deformation: Long-term deformation (creep) and short-term deformation (immediate deflection) can lead to serviceability issues, such as cracking or excessive movement.

Foundation Failure: This occurs when the soil beneath the structure is unable to adequately support the load of the structure, leading to settlement, tilting, or even collapse of the structure.

Progressive collapse: It also known as disproportionate collapse, refers to the phenomenon where a localized failure of a primary structural component leads to the collapse of adjoining members which, in turn, leads to additional collapse. Hence, the total damage is disproportionate to the original cause. A primary scenario that can cause progressive collapse in reinforced concrete structures is the sudden loss of a critical load-bearing element, such as a column or a key load-bearing wall. This can be due to extreme events such as explosions, fires, earthquakes, or even design and construction errors. Progressive collapse is particularly dangerous because it can lead to widespread structural damage and loss of life, even from relatively minor initial failures. For this reason, modern design codes often require that buildings be designed to mitigate the risk of progressive collapse [58, 59].

Each of these cases can be mitigated through proper design, high-quality construction methods, regular maintenance, and the use of appropriate materials.

9. Case studies: reinforced concrete building failures

Understanding the failures of the past is crucial for improving the design and construction practices of the future. This section presents several detailed case studies of some notable building failures involving reinforced concrete structures.

9.1 Failure of the L'Ambiance Plaza collapse, USA, 1987

The L'Ambiance Plaza was a 16-story residential project under construction in Bridgeport, Connecticut, when it collapsed in April 1987. The collapse resulted in the death of 28 construction workers, making it one of the worst disasters in modern U.S. construction history. The construction method employed was lift-slab, wherein

Figure 6.
The L'Ambiance plaza collapse [58].

the concrete slabs are cast on the ground and then lifted and placed on top of the columns. The subsequent investigation concluded that the collapse was due to the premature removal of the lifting apparatus before the slabs were adequately secured [58, 59]. The lessons learned from the L'Ambiance Plaza collapse led to increased scrutiny of the lift-slab construction method (**Figure 6**).

9.2 Failures in Elazığ earthquake, Turkey, 2020

In January 2020, a magnitude 6.7 earthquake struck the Elazığ province in eastern Turkey, causing significant damage and loss of life. Several reinforced concrete buildings in the region collapsed or were severely damaged, revealing deficiencies in their design and construction (**Figure 7**). Common issues such as soft-story mechanisms (where the ground floor is weaker and more flexible than the upper floors), inadequate confinement of concrete columns, and poor-quality concrete and reinforcement were discovered in the investigation [60].

The earthquake served as a grim reminder of the importance of proper seismic design and construction practices, particularly in regions with a high seismic risk. It also underscored the need for enhancing quality control in construction and enforcement of building codes to ensure the safety of the structures and the occupants (**Figure 7**).

9.3 Collapse of Champlain Towers South, USA, 2021

In June 2021, a 12-story condominium building in Surfside, Florida, partially collapsed, leading to a significant number of fatalities. The collapse of the Champlain Towers South is considered as one of the deadliest building failures in U.S. history (**Figure 8**). While the exact cause of the collapse is still under investigation at the time

DOI: http://dx.doi.org/10.5772/intechopen.1002247

Figure 7.
Damages to adjacent buildings during the Elazığ earthquake [60].

of writing, preliminary reports suggest that there were serious issues with the building's foundation and waterproofing, which led to significant corrosion of the reinforcement in the concrete slabs and columns [61, 62].

The collapse has prompted a reevaluation of building inspection and maintenance practices, particularly for older buildings situated in corrosive environments. It has also emphasized the importance of promptly addressing any identified structural issues to prevent catastrophic failures.

Figure 8.
Collapse of Champlain towers south [61].

9.4 Collapse of the CTV building, New Zealand, 2011

The CTV building was a six-story office block situated in Christchurch, New Zealand. During a 6.3 magnitude earthquake in February 2011, the building collapsed, causing 115 fatalities. The collapse was primarily attributed to design deficiencies in the building's shear walls and the poor layout of vertical structural elements. An eccentric building like CTV would move more at the corners than the middle in an earthquake, and this shape eccentricity has contributed to the building collapse [63, 64].

The CTV building collapse underscored the importance of rigorous seismic design and the potential for disproportionate collapse when key structural elements fail. It highlighted the necessity for robust building codes and standards that consider seismic events, ensuring that buildings can withstand such extreme forces (**Figure 9**).

9.5 Failures in the Kahramanmaraş earthquake, Turkey, 2023

April 28, 2023 an earthquake measuring 6.5 on the Richter scale hit the city of Kahramanmaraş in Turkey, and caused extensive damage to several structures, among which many reinforced concrete buildings.

The immediate post-event assessment estimated that approximately 120 reinforced concrete buildings in the city either collapsed or suffered severe structural damage that made them uninhabitable. Tremors were felt across the city following the earthquake, leading to more disintegrations and failures of particularly older structures constructed using reinforced concrete which failed to withstand the earthquake's intense shaking (**Figure 10**).

Investigations into the building failures revealed that several factors contributed to the failures, including among other factors: inadequate reinforcement detailing, poor concrete quality, and overall poor construction practices [65].

Figure 9.
Collapse of the CTV building, New Zealand [63, 64].

DOI: http://dx.doi.org/10.5772/intechopen.1002247

Figure 10.
A photo of the building damage from the 2023 Kahramanmaraş earthquake, Turkey [65].

9.6 Fire-induced partial collapse at the Faculty of Architecture Building at Delft University of Technology, the Netherlands, 2013

The degradation of reinforced concrete structures due to fire poses a substantial challenge in structural engineering. Under high-temperature conditions, the concrete experiences both physical and chemical modifications, which impair its mechanical characteristics and compromise the overall structural stability. The transformation of the concrete's inherent moisture into steam results in internal pressures that can provoke explosive spalling. This process not only unveils the underlying reinforcing steel but also diminishes the entire cross-section of the structure, thereby further undermining its load-bearing potential [66].

Moreover, the reinforcing steel within the concrete is prone to losing its strength and rigidity under high temperatures. This loss hampers the structure's resilience against deformation, steering it toward potential failure. As fire hazards continue to pose a threat to structures, it becomes imperative to understand the dynamics of fire-induced failure in reinforced concrete. This understanding will pave the way toward more fire-resistant materials and structural designs.

Although the complete structural collapse of high-rise buildings due to fire is an uncommon occurrence, large fires leading to localized structural damage or partial collapses are a more regular phenomenon. For instance, on May 13, 2008, a fire-induced partial collapse was reported at the Faculty of Architecture Building at Delft University of Technology in the Netherlands. The fire culminated in the structural collapse of a considerable portion of the building (see **Figure 11**) [67].

Figure 11.
Fire-induced partial collapse of the Faculty of Architecture Building at Delft University of Technology in the Netherlands [67].

9.7 Progressive collapse of the Ronan Point apartment tower collapse in London in 1968

One infamous example of progressive collapse is the Ronan Point apartment tower collapse in London in 1968. A small gas explosion in one of the apartments on the 18th floor led to the failure of a load-bearing wall, which triggered a progressive collapse of the entire corner of the building, resulting in the death of four people and injuries to 17 others. This incident led to significant changes in building codes in the UK and worldwide to prevent similar collapses (**Figure 12**) [69–70].

9.8 Progressive collapse of the Murrah Federal Building in Oklahoma City, USA, in 1995 due to a terrorist attack

Another example is the partial collapse of the Murrah Federal Building in Oklahoma City, USA, in 1995 due to a terrorist attack. It serves as a sobering reminder to structural engineers of the potential consequences of intentional acts of destruction. This tragic incident highlighted the critical need for structural designs that can withstand extreme events and mitigate the propagation of localized failures. The collapse mechanism observed in this case underscored the importance of considering robust security measures and incorporating progressive collapse resistance strategies in building designs. As structural engineers, it is our responsibility to continually improve our understanding of the dynamic behavior of structures under extreme loading conditions and to develop innovative design solutions that enhance the resilience and safety of buildings. The lessons learned from this event have since played a pivotal role in shaping security protocols, risk assessment methodologies,

Figure 12.
Progressive collapse of the Ronan point apartment tower collapse in London in 1968 [68].

and counterterrorism measures within the field of structural engineering, ensuring that our structures can better withstand intentional acts of destruction and protect the lives and well-being of occupants (**Figure 13**) [71].

The key strategy to mitigate progressive collapse is to provide alternative load paths in the structure. If one component fails, the load it was carrying is safely

Figure 13.
Progressive collapse of the Murrah Federal Building in Oklahoma City, USA, in 1995 due to a terrorist attack [71].

transferred to other, redundant elements. This is often achieved by designing the structure to be ductile, especially at the connections, and by adding reinforcements or using prestressed concrete to enhance structural integrity.

In summary, the threat of progressive collapse in reinforced concrete structures is a significant design consideration. It involves understanding potential weak points in a structure and designing them to either be robust enough to withstand abnormal loads or flexible enough to redistribute the loads to other elements should failure occur. Awareness and careful consideration of progressive collapse can result in safer, more resilient structures.

10. Future directions in reinforced concrete

The realm of reinforced concrete is perpetually in flux, guided by developments in material science, design methodologies, and novel construction technologies. This comprehensive section delves into potential future directions in the field, exploring advanced materials, digital technologies, sustainability efforts, and resilience-based design strategies.

10.1 Advanced materials

The future of reinforced concrete is intertwined with advancements in material science. The next generation of concrete and reinforcement materials promises superior properties and better structural integrity.

10.1.1 Ultra-high-performance concrete (UHPC)

Characterized by its superior properties compared to traditional concrete, UHPC offers exceptional strength and durability. Regular concrete usually has a compressive strength of about 3000 psi, but UHPC surpasses this sevenfold, with strengths exceeding 22,000 psi (150 MPa).

Furthermore, UHPC boasts excellent durability, with high resistance to corrosion, abrasion, and impact. However, it is important to note that the complex mix design and associated high cost currently restrict its widespread use.

10.1.2 Shape memory alloys (SMAs)

Shape Memory Alloys (SMAs) present a revolutionary concept for reinforced concrete. These materials possess the ability to regain their original shape after deformation—a property known as the shape memory effect (**Figure 14**) [72, 73].

The integration of SMAs into reinforced concrete structures holds immense promise for mitigating damage, improving structural integrity, and facilitating self-repair after calamitous events like earthquakes. This application remains largely experimental but offers considerable potential for future development [73].

10.2 Digital technology

The integration of digital technologies into the field of reinforced concrete provides valuable tools for more precise and efficient design and construction processes.

DOI: http://dx.doi.org/10.5772/intechopen.1002247

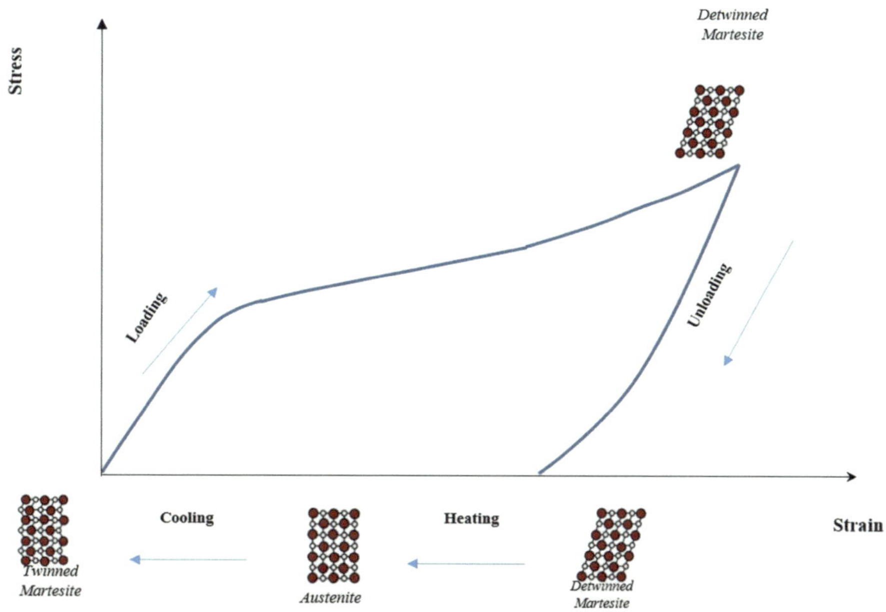

Figure 14.
Diagram illustrating the shape memory effect in SMAs [71].

10.2.1 Building information modeling (BIM)

Building Information Modeling (BIM) is an innovative method that harnesses digital technology to create, manage, and manipulate a virtual representation of a building's physical and functional characteristics. Within the context of reinforced concrete structures, BIM provides a platform to collaboratively explore, evaluate, and optimize design choices, thus ensuring the integrity of the structure.

BIM is more than just a detailed 3D model of the building; it is an intelligent information repository that includes data about every component of the structure. For reinforced concrete, this might include data about the concrete mix design, the size and placement of the reinforcement bars, and the expected load-bearing capacity. This comprehensive information allows engineers to perform detailed analyses and make informed decisions about the design and construction process.

Through BIM, architects, engineers, and contractors can collaborate effectively, visualizing and addressing potential structural, architectural, and construction issues early in the design phase. This collaborative approach results in better-coordinated projects, reducing the likelihood of costly errors and rework during construction.

In the maintenance phase, BIM can contribute to the structural integrity of reinforced concrete structures by providing a detailed, readily accessible database of the building's components. This can inform maintenance activities and provide valuable data in the event of a required intervention or rehabilitation process. BIM could even integrate with Structural Health Monitoring (SHM) systems, providing a platform to visualize, interpret, and act upon the data gathered by these systems [74].

In conclusion, BIM stands as a revolution in the field of reinforced concrete design and construction [74, 75]. It is not only a tool for creating detailed 3D models but also a comprehensive database, a collaborative platform, and an advanced analysis environment. By reducing errors, enhancing collaboration, and enabling sophisticated analyses, BIM significantly contributes to reinforced concrete structural integrity.

10.2.2 Artificial intelligence: from expert systems to machine learning

Artificial Intelligence (AI) has significantly evolved in its application to the field of structural engineering, specifically for reinforced concrete structures. This progression can be traced from the early use of expert systems, followed by the adoption of neural networks, and culminating with the latest innovations involving machine learning. Each phase of this evolution has brought about remarkable changes and improvements in how we understand, design, analyze, and manage reinforced concrete structures.

In the early days of AI, expert systems were developed to harness the knowledge and expertise of human experts in a rule-based system. These systems were used to guide less experienced engineers in making design and construction decisions related to reinforced concrete structures. Expert system can be used to provide recommendations for appropriate concrete mix design or reinforcement layout based on input parameters such as load requirements, environmental conditions, and project constraints. Examples of early expert system applications in structural concrete include: A system for concrete mix design in normal and hot climates an expert system for the preliminary design of earthquake-resistant buildings [76, 77], and a knowledge-based approach for the maintenance of reinforced concrete buildings in hot climates [78], and an advisory graphical system for crack diagnosis and assessment (**Figure 15**) [25].

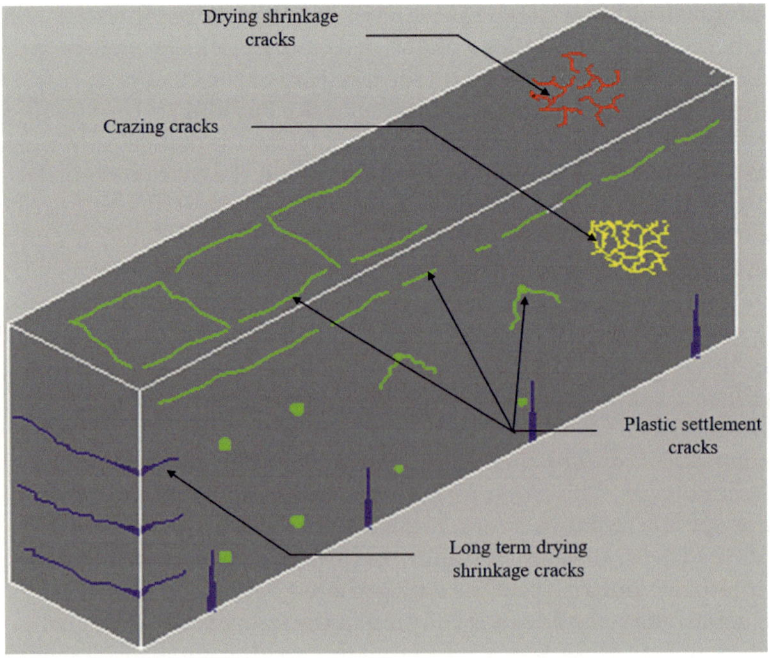

Figure 15.
Sample of an early AI expert system for the diagnosis of cracks in reinforced concrete [25].

With the advent of neural networks, AI in the field of reinforced concrete structures took a leap forward. Neural networks, inspired by the human brain's workings, enabled more sophisticated analysis of complex structures and phenomena. Neural network come up with solutions based on its training [79].

In the most recent phase of AI evolution, machine learning has emerged as a powerful tool for reinforced concrete structures. Machine learning algorithms, unlike traditional rule-based systems, improve their performance iteratively as they are exposed to more data. They have found applications in a variety of areas in structural concrete.

In summary, the use of AI in reinforced concrete structures has grown from simplistic rule-based expert systems to sophisticated machine learning models capable of learning from data and improving their performance over time. This evolution continues to open new possibilities for the design, analysis, and management of reinforced concrete structures.

10.3 Sustainable construction

Sustainability has become a key driving factor in the construction industry, urging professionals to devise methods to create structures that minimize environmental impact while maximizing efficiency and performance. The realm of reinforced concrete is no exception, as researchers and professionals in the field actively seek to imbue sustainability into every aspect of the process, from material selection to production methods and even end-of-life disposal.

10.3.1 Green concrete

At the forefront of sustainable concrete construction is the concept of Green Concrete. More than a specific type of concrete, Green Concrete encompasses an approach to concrete construction that prioritizes environmental sustainability. The development and use of Green Concrete involve careful selection of materials and optimized production processes aimed at minimizing environmental impact [80, 81].

Material selection often involves the use of recycled or waste materials as part of the aggregate or as supplementary cementitious materials (SCMs). SCMs could include by-products from other industries, such as fly ash from power plants or slag from metal processing industries. Utilizing these materials not only reduces the need for raw materials but also prevents these by-products from ending up in landfills.

Reducing cement content is another strategy employed in Green Concrete, as the production of cement is a significant contributor to global CO2 emissions. Instead, other binding materials like lime or SCMs may be used. Research and development are ongoing in the field to find alternative binding materials that can match or even surpass the performance of traditional cement.

Green Concrete can also involve the use of carbon capture technologies during cement production. These technologies can significantly reduce the carbon footprint of cement production, making the entire process more sustainable.

10.3.2 Life-cycle assessment (LCA)

LCA is a method used to evaluate the environmental impacts of a product or system from its inception to its end of life. This includes the extraction and processing of raw materials, the manufacturing process, the use and maintenance of the product, and its final disposal or recycling [65].

In the context of reinforced concrete structures, an LCA would include the extraction and processing of raw materials like aggregate and cement, the production and transport of the concrete, the construction process, the service life of the structure (including maintenance and potential repairs), and the eventual demolition and disposal or recycling of the structure.

By employing LCA in the design and construction process, professionals can identify areas where environmental impact can be reduced. This might involve choosing more sustainable materials, optimizing production and construction processes, extending the service life of the structure, or planning for efficient demolition and recycling at the end of the structure's life [81].

10.3.3 Sustainable use of water in concrete production

Water plays a vital role in the production of concrete, but the construction industry is one of the largest consumers of freshwater resources. Recognizing this, sustainable water management in concrete production is becoming a priority [82].

Recycling and reusing water from concrete production, known as wash water, is a key strategy in this area. Wash water, typically high in pH and containing trace amounts of cement, can potentially be reused in the concrete production process itself, thereby reducing the demand for freshwater.

10.3.4 Reducing, reusing, and recycling concrete waste

Construction and demolition waste constitutes a significant portion of total solid waste produced globally, and concrete waste forms a large part of this. Sustainable management of concrete waste is thus a crucial aspect of green construction [80].

Concrete structures at the end of their lifecycle need to be effectively managed. One such strategy is the recycling of concrete waste into Recycled Concrete Aggregates (RCA) [42, 44], which can be used in the production of new concrete. While there are challenges to overcome regarding the potentially variable properties of RCA, it offers a valuable route to reducing the demand for virgin aggregate and reducing landfill waste.

Finally, reducing concrete waste in the first place is paramount. This can be achieved through accurate calculation and optimization of concrete quantities required for a job, minimizing over-ordering, and effective management of leftover concrete. Employing these strategies helps move the reinforced concrete construction industry toward a more circular economy model, promoting sustainability.

10.4 Resilience-based design

The concept of resilience-based design is a holistic approach gaining significant traction in the realm of structural engineering, especially within the field of reinforced concrete. Resilience-based design goes beyond the traditional focus on the safety and serviceability of structures, aiming to create systems that not only withstand but also recover quickly from extreme events like earthquakes, floods, or hurricanes.

Resilience-based design takes into consideration the broader societal, economic, and environmental implications of structural failure. This includes the direct and indirect costs associated with structural repair and replacement, as well as the economic impact of disruptions to services and facilities housed within the structures. It also takes into account potential environmental impacts associated with structural damage or failure [81].

One important principle of resilience-based design is the notion of designing structures to exceed the minimum standards specified by codes. This "code-plus" approach involves designing structures to perform well under loads exceeding those prescribed by design codes, essentially preparing structures for worst-case scenarios.

In addition to this, resilience-based design involves proactive planning for the entire lifecycle of the structure. This includes careful consideration of the materials used, the construction methods, operation and maintenance, and even end-of-life deconstruction or recycling of the structure. By considering the entire lifecycle of the structure, the resilience-based design approach aims to reduce the overall environmental impact and promote sustainability.

A major aspect of resilience-based design is the integration of Structural Health Monitoring (SHM) systems. SHM involves the incorporation of advanced sensors and data analysis algorithms that can monitor the structural performance of the building over time. This allows for early detection of potential issues, enabling timely intervention and repair, thus minimizing downtime and disruption in the event of an extreme event [38].

Furthermore, the use of advanced modeling and simulation tools is a crucial part of resilience-based design. These tools enable engineers to predict how structures will respond under various scenarios, allowing for design optimization and proactive preparation for potential events. This includes seismic modeling for earthquake-prone regions, hydrodynamic modeling for structures exposed to potential flooding, and even aerodynamic modeling for buildings exposed to high wind loads.

11. Conclusion

The dynamic nature of reinforced concrete as a material, and its integral role in the construction industry, is indisputable. This chapter has provided an overview of the history of reinforced concrete, its fundamental principles, the progress made in the understanding of its behavior under various loads, and recent advances that have led to the development of stronger, more durable, and more versatile concrete structures.

Significant advancements have been made in the types of concrete available, including High-Performance Concrete, Fiber-Reinforced Concrete, and Self-Compacting Concrete. These developments allowed for more robust and efficient designs, reduced maintenance costs, and longer service life. Other recent advances in reinforced concrete look promising, including advanced materials like ultra-high-performance concrete and shape memory alloys, and smart concrete.

The chapter also highlighted some notable reinforced concrete building failures. These case studies serve as reminders of the consequences of inadequate design, poor construction practices, and failure to adhere to building codes. The lessons learned from these failures are invaluable in promoting safer construction practices and informing revisions to building codes.

The future of reinforced concrete is one of exciting possibilities, driven by the relentless pursuit of knowledge and the ambition to build structures that are not only strong and durable, but also contribute positively to our societies and the environment.

In conclusion, the field of reinforced concrete will continue to evolve, driven by innovation, research, and the imperative to learn from past failures. As we move forward, it is crucial that we continue to push the boundaries of what is possible, while always prioritizing safety, reliability, structural integrity, and sustainability.

Author details

Abdul Rahim Sabouni
Dalhousie University, Halifax, Nova Scotia, Canada

*Address all correspondence to: sabouni@gmail.com

References

[1] Nilson AH, Darwin D, Dolan CW. Design of Concrete Structures. 15th ed. USA: McGraw-Hill; 2016. p. 803

[2] McCormac JC, Brown RH. Design of Reinforced Concrete. 10th ed. USA: Wiley; 2017. p. 944

[3] Wight JK, Kong FK. Reinforced Concrete: Mechanics and Design. 7th ed. UK: Pearson Education; 2015. p. 672

[4] Fanella DA. Reinforced Concrete Structures: Analysis and Design. 2nd ed. USA: CRC Press; 2017. p. 480

[5] MacGinley TJ, Choo BS. Reinforced Concrete: Design Theory and Examples. 2nd ed. USA: Spon Press; 2008. p. 408

[6] Park R, Paulay T. Reinforced Concrete Structures. USA: John Wiley & Sons; 1975. 769 p. DOI: 10.1002/9780470172834.index

[7] ACI Committee 318. Building Code Requirements for Structural Concrete (ACI 318-14) and Commentary (ACI 318R-14). USA: American Concrete Institute; 2014

[8] Eurocode 2. Design of Concrete Structures - Part 1-1: General Rules and Rules for Buildings. Belgium: European Committee for Standardization; 2004

[9] Kind-Barkauskas F. The development of concrete technology. In: Concrete Construction Manual. München: DETAIL; 2002. pp. 7-17. DOI: 10.11129/detail.9783955531638.7

[10] Helbig T, Kleiser M, Krontal L. Bridges for slow-moving traffic. In: Bridges. München: DETAIL; 2021. pp. 14-21. DOI: 10.11129/9783955535643-004

[11] McBeth DG, Hennebique F, Mouchel LG. Francois Hennebique, (1842-1921), reinforced concrete Pioneer. Proceedings of the Institution of Civil Engineers - Civil Engineering. 1998;**126**(2). DOI: 10.1680/icien.1998.30436

[12] Zastavni D et al. Load path and prestressing in conceptual design related to Maillart's Vessy Bridge. In: Brasil RMLRF, Pauletti MO, editors. IASS-SLTE 2014 Symposium Shells, Membranes and Spatial Structures: Footprints; 2014

[13] May R. Architects and Engineers: Modes of Cooperation in the Interwar Period, 1919-1939. Berlin, Boston: Birkhäuser; 2022. DOI: 10.1515/9783035623260

[14] Bruun E, Maillart R. The evolution of reinforced concrete bridge forms, conference. In: The 9th International Conference on Short and Medium Span Bridges. Calgary: SciHi; 2014. pp. 1-10

[15] Fischer P, Gendolla T. Robert Maillart's key methods from the Salginatobel bridge design process. Journal of the International Association for shell and Spatial Structures. 2014;**53**:39-47

[16] Nawy EG. Prestressed Concrete: A Fundamental Approach. 6th ed. UK: Pearson Education; 2016

[17] ACI Committee 212. Chemical Admixtures for Concrete. USA: American Concrete Institute; 2010

[18] Ashour AF, Abdelrahman A. Effect of fiber types on mechanical properties of reinforced concrete. USA: Journal of Engineering Research

and Applications. 2018;**8**(9):1-13. DOI: 10.9790/9622-080901113

[19] Hamid A, Sabouni AR, Issa M; Mukhtar A. Use of fiber reinforced plastics in reinforced concrete slabs. In: Proceedings of the Sixth Arab Structural Engineering Conference, October 21-24. Vol. 2. Damascus, Syria: Damascus University; 1995. pp. 237-248

[20] Bilek V, Hurta J, Done P, Zidek L. Comparison of properties of concretes with different types and dosages of fibers. In: Proceedings of the 2nd Czech-China Scientific Conference 2016. London, UK: InTech; 2017. DOI: 10.5772/66804

[21] Zhang MH, Malhotra VM. High-Performance Concrete. UK: Taylor & Francis; 2004

[22] Kim YJ, Kim JH. Early-Age Strength of Ultra-High Performance Concrete in Various Curing Conditions. USA; 2015. Available from: https://www.ncbi.nlm.nih.gov/pmc/articles/PMC5455500/

[23] Self-healing Concrete to build for the Ages. Restoration of Buildings and Monuments. Self-healing Concrete to build for the Ages, Germany; 2013;**19**(1):49-50. DOI: 10.1515/rbm-2013-6573

[24] Kashef M, Sabouni AR. Nanotechnology and the building industry. In: Paper Presented at the International Conference on Nanotechnology: Fundamentals and Applications. Ottawa, Ontario, Canada; 2010. pp. 551-558

[25] Sherif AM, Sabouni AR, Berrais A, Alshamsi AM. Advisory graphical system for crack diagnosis and assessment. In: Proceedings of the Second International Conference in Civil Engineering on Computer Applications, Research and Practice. Bahrain: Bahrain University; April 6-8, 1996. pp. 31-40

[26] Sabouni AR. Expert system methodology for deterioration of reinforced concrete structures. In: Proceedings of the Fifteenth Australian Conference on the Mechanics of Materials and Structural Mechanics (ACMSM). Melbourne, Australia: A. A. Balkema Publishing; 8-10 December 1997. pp. 251-256

[27] Zhiqiang D, Gang W, Jinlong L. Experimental study on the durability of FRP bars reinforced concrete beams in simulated ocean environment. Science and Engineering of Composite Materials. 2018;**25**(6):1123-1134. DOI: 10.1515/secm-2017-0237

[28] Al Martini S, Al Mzayyen N, Sabouni R, Shahria A. Investigation on Bond Performance between Basalt FRP Rebars and Recycled Aggregate Concrete. Vol. 248. Canada: Lecture Notes in Civil Engineering; 2023. pp. 695-705

[29] Casandjian C, Challamel N, Lanos C, Hellesland J. Concepts for the Design at Ultimate Limit State (ULS). In: Reinforced Concrete Beams, Columns and Frames (Chapter 3). 2013. DOI: 10.1002/9781118639511.ch3

[30] Bhattacharya B, Nagesh R. The serviceability limit states in reinforced concrete design. International Journal of Engineering Sciences & Emerging Technologies. 2013;**5**(2):101-112

[31] Fajfar P, Krawinkler H. Performance-Based Seismic Design - Concepts and Implementation. USA: Pacific Earthquake Engineering Research Center; 2005

[32] Rana S, Gupta R. Performance based seismic design of reinforced concrete

buildings for risk reduction against earthquakes forces-an overview. Journal of Civil Engineering and Environmental Technology. 2017;**4**(3):253-257

[33] Al-Shaibani A, Sabouni R. Sustainability considerations of concrete mixes incorporating slag as partial replacement of cement in concrete in the UAE. Key Engineering Materials. 2016;**677**:217-223

[34] Al Martini S, Sabouni R, El-Sheikh A. Fresh, durability and mechanical behavior of self consolidating concrete incorporating Fly ash and admixture under hot weather. Key Engineering Materials. 2021;**904**:453-457

[35] Watts A. Fibre reinforced concrete. In: Modern Construction Handbook. Berlin, Boston: Birkhäuser; 2023. pp. 52-55. DOI: 10.1515/9783035624960-019

[36] Gołaszewski J, Cygan G, Ponikiewski T, Gołaszewska M. Cements with calcareous fly ash as component of low clinker eco-self compacting concrete. Open Engineering. 2019;**9**(1):196-201. DOI: 10.1515/eng-2019-0024

[37] Yan Y, Xing Z, Chen X, Xie Z, Zhang J, Chen Y. Axial compression performance of CFST columns reinforced by ultra-high-performance nano-concrete under long-term loading. Nanotechnology Reviews. 2023;**12**(1):20220537. DOI: 10.1515/ntrev-2022-0537

[38] Song G. Structural Health Monitoring (SHM) of Civil Structures. Switzerland: MDPI AG; 2018

[39] Li VC. Engineered Cementitious Composites (ECC): Bendable Concrete for Sustainable and Resilient Infrastructure. Germany: Springer; 2019

[40] Sabouni R, Abdulhameed H. Optimization of sustainable concrete mixes containing binary and ternary blends. Key Engineering Materials. 2021;**834**:131-141

[41] Sabouni R, Abdulhameed H. Investigation on performance of concrete incorporating silica fume and local UAE materials. Key Engineering Materials. 2019;**803**:222-227

[42] Al Martini A, Sabouni R, Khartabil A, Tadesse G, Wakjira M, Alam S. Development and strength prediction of sustainable concrete having binary and ternary cementitious blends and incorporating recycled aggregates from demolished UAE buildings: Experimental and machine learning-based studies. Construction and Building Materials. 2023;**2023**:380

[43] Murthi P, Awoyera P, Selvaraj P, Dharsana D, Gobinath R. Using silica mineral waste as aggregate in a green high strength concrete: workability, strength, failure mode, and morphology assessment. Australian Journal of Civil Engineering. 2018:1-7. DOI: 10.1080/14488353.2018.1472539

[44] Martini S, Khartabil A, Neithalath N. Rheological properties of sustainable concrete incorporating recycled aggregates and supplementary cementitious materials. ACI, I Theme: Advances in Rheology and AM. 2021;**118**(6):241-255

[45] El Naggar A, El Naggar H, Sadeghian P. Simplified material model for concrete containing high-content of Tire-derived coarse aggregate under compression loading. Canadian Journal of Civil Engineering. 2020;**48**:912-924. DOI: 10.1139/cjce-2019-0626

[46] Lofinnade Oluwarotimi M et al. Strength and microstructure

of eco-concrete produced using waste glass as partial and complete replacement for sand. Cogent Engineering. 2018;**5**(1):1483860. DOI: 10.1080/23311916.2018.1483860

[47] Martini S, Sabouni R, Khatarbil A. Evaluation of thermal conductivity of sustainable concrete having supplementary cementitious materials (SCMs) and recycled aggregate (RCA) using needle probe test. Sustainability. 2023;**15**(1):109

[48] Kloft H et al. Reinforcement strategies for 3D-concrete-printing. Civil Engineering Design. 2020;**2**:131-139. DOI: 10.1002/cend.202000022

[49] Han B, Wang Y, Dong S, Zhang L, Ding S, Yu X, et al. Smart concretes and structures: A review. Journal of Intelligent Material Systems and Structures. 2015;**26**(11):1303-1345. DOI: 10.1177/1045389X15586452

[50] Available from: https://www.civil-engg-world.com/2020/01/07/reliability-of-rebound-hammer-test-on-concrete/concrete/

[51] Available from: https://www.screeningeagle.com/en/products/pundit-200

[52] Barrile V, Pucinotti R. Application of radar technology to reinforced concrete structures: A case study. NDT E Int. 2005;**38**:596-604

[53] Barnes CL, Trottier JF, Forgeron D. Improved concrete bridge deck evaluation using GPR by accounting for signal depth–amplitude effects. NDT E Int. 2008;**41**:427-433

[54] Li X, Xiao Y, Guo H, et al. A BIM based approach for structural health monitoring of bridges. KSCE J Civil Eng. 2022;**26**:155-165. DOI: 10.1007/s12205-021-2040-3

[55] Malhotra VM. Handbook on Nondestructive Testing of Concrete. Second ed. USA: CRC Press; 2004

[56] Werkle H. Finite Elements in Structural Analysis: Theoretical Concepts and Modeling Procedures in Statics and Dynamics of Structures. Germany: Springer International Publishing; 2021

[57] Sabouni A R, El-Zanaty A, Igraffea A. Finite element discretization of mixed-mode fracture. Proceedings of the Ninth Conference on Electronic Computations, American Society of Civil Engineers, Alabama, USA, 1986, p. 257-267

[58] Martin R, Delatte N. Another look at the L'Ambiance plaza collapse. Journal of Performance of Constructed Facilities. 2000;**14**(4):160-165

[59] Staff reports, CTpost. What to know about the 1987 L'Ambiance Plaza collapse in Bridgeport. Connecticut Post. Available from: https://www.ctpost.com/news/article/l-ambiance-plaza-collapse-bridgeport-17119543.php [Accessed: April 24, 2023]

[60] Sayın E, Yön B, Onat O, et al. Sivrice-Elazığ, Turkey earthquake: Geotechnical evaluation and performance of structures. Bulletin of Earthquake Engineering. 2020;**19**(2021):657-684. DOI: 10.1007/s10518-020-01018-4

[61] NIST. Champlain Towers South Collapse, Disaster and Failure Studies Program. 2021. Available from: https://www.nist.gov/disaster-failure-studies/champlain-towers-south-collapse-ncst-investigation [Accessed: April 5, 2022].

[62] Simons R, Xiao Y, Evenchik A, Barreto A. Champlain towers south

collapse: Frequency, governance and liability issues. Journal of Sustainable Real Estate. 2022;**14**(1):57-74. DOI: 10.1080/19498276.2022.2104346

[63] Seibel W. Erosion of professional integrity: The collapse of the Canterbury TV building in Christchurch. In: Collapsing, Structures and Public Mismanagement. Cham: Palgrave Macmillan; 2022. DOI: 10.1007/978-3-030-67818-0_4

[64] Royal Commission on the Canterbury Earthquakes. Report on the Collapse of the CTV Building, Christchurch. New Zealand, Wellington: NZ Government website; 2012. p. 118

[65] Ministry of Environment and Urbanization. Earthquake report: Kahramanmaraş, Turkey, 2023. Ankara, Turkey: Support To Life; 2023. p. 143

[66] Li G-Q, Zhang C, Jiang J. A review on fire safety engineering: Key issues for high-rise buildings. International Journal of High-Rise Buildings. 2018;7(4):265-285. DOI: 10.21022/IJHRB.2018.7.4.265

[67] Engelhard MD et al. Observations from the Fire and Collapse of the Faculty of Architecture Building. Delft, Netherlands: Delft University of Technology; 2013

[68] Designing Buildings. Ronan Point. The Construction WIKI. Available from: https://www.designingbuildings.co.uk/wiki/Ronan_Point.

[69] Rouse, James W, Delatte NJ. Lessons from the progressive collapse of the Ronan point apartment tower. In: Harris JR, Wood RL, editors. Proceedings of the 3rd ASCE Forensics Congress. American Society of Civil Engineers; 2003. pp. 1-10

[70] Birch Jules. "Ronan Point 50 Years On: The Worrying Legacy of a Disaster." Inside Housing, Comment, 2018. Available from: https://www.insidehousing.co.uk/comment/ronan-point-50-years-on-the-worrying-legacy-of-a-disaster-56236#gallery-1

[71] Available from: https://cdn.britannica.com/21/74221-050-70FD60D9/remains-Oklahoma-City-Alfred-P-Murrah-Federal-April-19-1995.jpg

[72] Alam MS, Youssef MA, Nehdi M. Utilizing shape memory alloys to enhance the performance and safety of civil infrastructure: A review. Canadian Journal of Civil Engineering. 2007;**34**:1075-1086. DOI: 10.1139/l07-038

[73] Shameli-Derakhshan T, BT, S, Jarrahi H. An overview on active confinement of concrete column and piers using SMAs. In: Paper Presented at the 2nd International Conference on Researches in Science. Istanbul, Turkey; February 2017

[74] Sadek K, El-Bastawissi I, Raslan R, Sayary S. Impact of BIM on building design. BAU Journal - Creative Sustainable Development. 2019;**1**(1):1

[75] Azhar S. Building information modeling (BIM): Trends, benefits, risks, and challenges for the AEC industry. Leadership & Management in Engineering. 2011;**11**:241-252. DOI: 10.1061/(ASCE)LM.1943-5630.0000127

[76] Sabouni AR, Alshamsi AM, Sherif AM. An expert system approach for concrete mix design in normal and hot climates. In: Proceedings of the Second International Conference in Civil Engineering on Computer Applications, Research and Practice, Bahrain

University, Bahrain, April 6-8. 1996. pp. 41-48

[77] Sabouni AR. An expert system for the preliminary design of earthquake resistant buildings. In: Proceedings of the Second Congress on Computing in Civil Engineering, ASCE, Atlanta, Georgia, USA, June 5-7. 1995. pp. 1188-1195

[78] Sabouni AR. Knowledge-based approach for maintenance of reinforced concrete buildings in hot climates. In: Proceedings of the Second International Conference on Concrete under Severe Conditions-Environment and Loading, CONSEC'98, June 21-24 1998. Vol. 2. Tromso, Norway; 1998. pp. 1209-1217

[79] Hadi MNS. Neural networks applications in concrete structures. Computers & Structures. 2003;**81**(6):373-381. DOI: 10.1016/S0045-7949(02)00451-0

[80] Sureshkumar, et al. A review of green concrete studies and discussion on various recycled materials. 2019. Available from Research Gate website: https://www.researchgate.net/publication/351149676_Green_Concrete_-_A_ReviewA Review on Green Concrete. Available from Research Gate website: https://www.researchgate.net/publication/281774385_A_Review_on_Green_Concrete

[81] Zeug W, Bezama A, Thrän D. A framework for implementing holistic and integrated life cycle sustainability assessment of regional bioeconomy. International Journal of Life Cycle Assessment. 2021;**26**:1998-2023. DOI: 10.1007/s11367-021-01983-1

[82] Ghotbi A. Resilience-Based Seismic Evaluation and Design of Reinforced Concrete Structures. [Thesis]. Los Angeles: University of California; 2018

Chapter 2

Proposing a Design Model for Determining Flexural Bearing Capacity of RC Beams Reinforced by Steel with Reduced Modulus of Elasticity

Sanin Dzidic and Aldin Mahmutovic

Abstract

The design of concrete structures and elements in Europe and wider is conducted according to EN 1992-1-1. Among other design assumptions, the Eurocode 2 assumes the design value of the modulus of elasticity E_s of reinforcing steel to be 200 GPa. However, what happens in the RC beam if the actual modulus of elasticity is significantly reduced. Does it affect the flexural bearing capacity of RC beam and to what extent? Another logical question is how to determine the actual flexural bearing capacity of the RC beam reinforced with reinforcing steel with a reduced modulus of elasticity and which design model to use for such determination. This study tries to answer such questions using an experimental approach and assumed calculation model with a comparison of experimental and calculation results. The experimental research from this showed that test RC beams reinforced by steel with reduced modulus of elasticity have significantly reduced flexural capacity in comparison with the designed flexural capacity of beams reinforced by steel with "normal" modulus of elasticity. In this regard, it is recommended to test the mechanical properties of the steel reinforcement prior to the installation at the site and not to rely on the producer's factory production control certificate only. Additional issues considered in this research are observed effects of the reduced modulus of elasticity of reinforcing bars to Serviceability Limit States (stress limitation, crack width, and deflection control). Answers to such questions can inform decisions if retrofit is needed, is it feasible and if yes—which retrofit method to be used. This study does not discuss the reasons for the reduced modulus of elasticity in reinforcing steel.

Keywords: RC beams, reinforcing steel, modulus of elasticity, flexural bearing capacity, failure mode

1. Introduction

Reinforced concrete is a material widely used for construction all over the world, and there is practically no serious structure, building, or facility that has not been built in reinforced concrete as a whole or in part. Design of load-bearing structures is generally based on the strength of materials, where the homogeneity and isotropicity of the material are assumed. However, concrete is anything but, neither homogeneous nor isotropic. For this reason, the design of concrete structures is so complex.

Concrete is essentially a composite material, and even when reinforced with steel reinforcement with the basic task of taking over the tensile forces in the reinforced concrete element, that compositeness becomes even more pronounced. Each of the concrete ingredients and materials used contributes to the behavior of materials, elements, or the entire structure.

In this sense, the quality control of concrete, both in production and on the construction site itself, in accordance with design assumptions, is well developed both through regulations and standards, as well as their application in a large part of the world.

However, the steel reinforcement is purchased locally on the construction materials market and practitioners are usually satisfied with the manufacturer's certificates of the steel reinforcement without any further control of steel reinforcement and is installed in accordance with the reinforcement plans from the structural design. However, let us ask a hypothetical question—what happens to the element and the entire reinforced concrete structure if the steel reinforcement does not correspond to the assumptions from the standards or design? What serviceability issues can arise in the case of installation of inappropriate steel reinforcement? Can this cause consequences for the structure, building, or facility, are the deformations in accordance with the expected, is the load-bearing capacity endangered? Can there be serious consequences that will affect the load-bearing capacity and serviceability of the structure? Will it be necessary to retrofit the structure, is it even possible, in what way and how much will it cost? Can the building collapse and under what conditions lead to catastrophic consequences? And finally, the question—whose responsibility, is it? On the other hand, with modern methods of retrofit and strengthening of concrete structures, it is possible to significantly increase the load-bearing capacity of concrete structures by using state-of-the-art materials with respect to all technological processes for their application [1].

This research represents a contribution to the eventual elaboration of the mentioned aspects in the previous paragraph. Essentially, this research deals with the consequences of reinforced concrete construction if steel reinforcement with a reduced modulus of elasticity is installed in an element or structure.

In relation to the modulus of elasticity, the current practice is to assume to be about 200 GPa for all steel grades. However, tensile tests of these steels have consistently shown that the modulus of elasticity varies with grade steel and thickness [2].

There are several methods for determining the modulus of elasticity. The most known methods are mechanical (static and dynamic), acoustic, ultrasonic, resonant, optic, etc. [3–6]. Mechanical methods are the most compatible for determining the elastic modulus of thin materials such as rods, wires, and fibers [3]. "The mechanically measured Young's modulus of metals is consistently lower than the physically measured one, particularly after plastic straining. Furthermore, the nominally elastic loading and unloading behavior is not linear; it shows significant curvature and hysteresis. While many reports of this so-called "modulus effect" have appeared, the consistency of the behavior among grades of steel, or within a single grade produced by alternate methods and suppliers, is unknown" [7].

Lems [8] concluded that the mechanically measured Young's modulus is reduced by plastic straining. The study from Morestin and Boivin [9] showed that materials degraded up to approximately 20% during the first 5% of plastic strain and using the mechanically measured Young's modulus value. Further research showed similar degradation magnitudes, depending on the material: up to 30% for mild steels [10] and 20% for high-strength steels [10, 11]. Numerous tensile tests of different grade of steel and thickness revealed that the current practice of assuming the same modulus of elasticity value of 200 GPa for all grades and thicknesses of steel may be inaccurate [2].

With the increasing use of a range of high-strength steels and thin steels in the building and construction industry, a good knowledge of the basic mechanical properties of these steels, namely, the yield and tensile strengths, the modulus of elasticity and ductility parameters is needed [2].

The mechanical properties of steel rebars affect the load-bearing capacity of a particular structure [12]. However, there are no or very limited number of theoretical or experimental investigations related to how reduced modulus of elasticity affects the load-bearing capacity of the RC elements. This research is a contribution to this gap.

2. Research methodology

Experimental comparative research methodology was used in this research. It aimed to determine the effects of the steel reinforcement with reduced modulus of elasticity to the flexural capacity of the RC beams. The comparison was made to the design model according to EN 1992-1-1 [13] with steel reinforcement with a specified modulus of elasticity of 200 GPa.

There are two considered research questions in this study. The first one is does the reduced modulus of elasticity of steel reinforcement affect the flexural and load-bearing capacity of RC beams and how? The second question relates to the failure mode of RC beams reinforced by steel reinforcement with reduced modulus of elasticity.

The findings on these research questions can inform the necessity for further investigation on effects of altered mechanical properties of the applied materials to the behavior of load-bearing elements and structures in serviceability and ultimate limit states, as well as necessity for retrofitting.

3. Small-scale experimental setup

The goal of the experimental analysis in this research is to compare the actual flexural capacity and behavior of a reinforced concrete beam in the ultimate limit state, which is reinforced with steel reinforcement with a reduced modulus of elasticity compared to a reinforced concrete beam reinforced with steel reinforcement with a standard modulus of elasticity according to EN 1992-1-1 [13].

3.1 Test samples

The subject of the experimental test is a series with three RC beam samples. The dimensions of the beam were chosen so that they could be adapted to the capabilities and equipment of the laboratory, so it is a small-scale experimental analysis.

The experimental series of beams consists of three reinforced concrete beams with a width of 10 cm, a height of 20 cm, and a length of 140 cm. They are made of

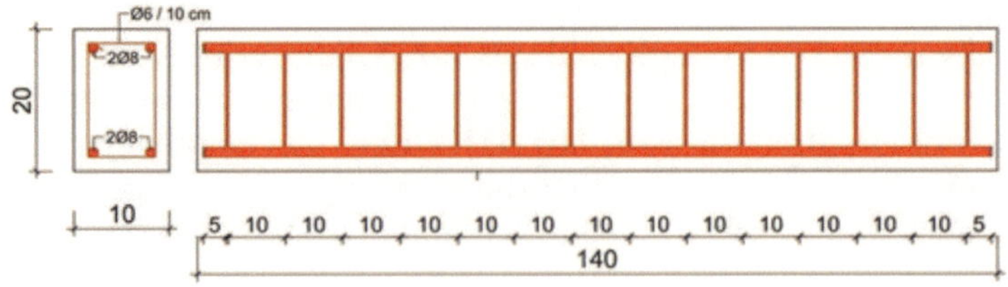

Figure 1.
Dimensions and reinforcement of test samples (cm).

concrete C 25/30 and reinforced with ±2 Ø 8 reinforcing steel f_{yk} = 500 MPa (**Figure 1**).

It should also be noted that the thickness of the concrete cover on the test beams is 1.0 cm. According to EN 1992-1-1 [13] and EN 1992-1-2 [14], this thickness of the concrete cover is not adequate for beams, but it was applied in this research in order to adapt the dimensions of the test beams to the capabilities of the equipment in the laboratory, and in order to increase the flexural capacity of the test samples.

3.2 Production of formwork for test samples

The formwork for the preparation of the samples was made according to **Figure 2**. Timber formwork consists of three elements:

- Formwork bottom 2.7 cm thick, dimensions 150 cm × 50 cm;
- Back sides made of planks 2.4 cm thick, 20 cm deep; and
- Protection against buckling and dimensional stability of the sides during casting of concrete- lath 3 cm × 5 cm.

According to the formwork plan, the formwork was produced and is displayed in **Figure 3**.

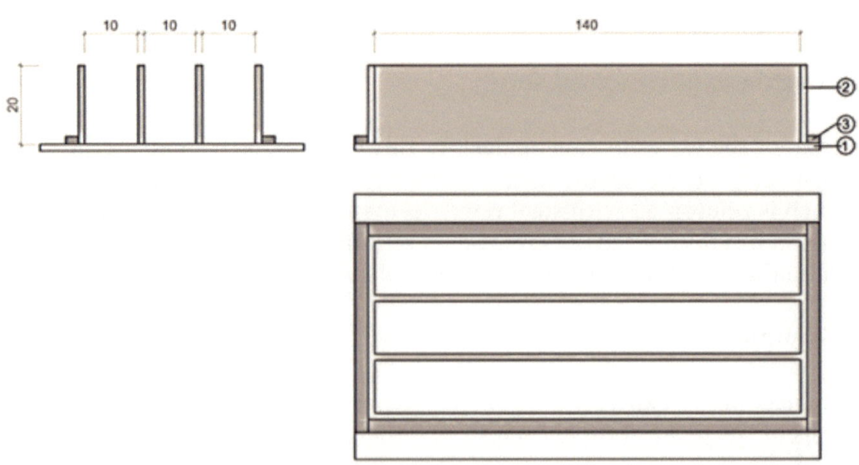

Figure 2.
Test samples formwork plan in cm.

Figure 3.
Test samples formwork.

3.3 Materials for test samples

According to the experiment design, it was envisioned that test samples will be made from concrete C 25/30. In order to prove the achieved class of concrete, three cube-shaped test samples of 15 × 15 × 15 cm/cm/cm were taken for compression strength testing. After 24 hours of concrete casting, the samples were taken out of the mold and immersed in water for conditioning, where they remained until the compression testing (**Table 1**).

EN 206 [15], Appendix B, Table B1 requires that each individual f_{ci} result must be greater than or equal to $f_{ck} - 4$, thus:

- Test Sample 1 - f_{c1} = 31.80 MPa ≥ (30.0 - 4.0) = 26.0 MPa;
- Test Sample 2 - f_{c2} = 29.60 MPa ≥ (30.0 - 4.0) = 26.0 MPa; and
- Test Sample 3 - f_{c3} = 31.70 MPa ≥ (30.0 - 4.0) = 26.0 MPa.

In addition, the mean value of the results (f_{cm}) must be equal to or greater than f_{ck} + 1, so:

- Mean value f_{cm} = 31.03 MPa ≥ (30.0 + 1.0) = 31.0 MPa.

Therefore, according to EN 206 [15] and the criterion of identity, the concrete from which the samples of reinforced concrete beams were made is classified as concrete class C 25/30.

Designation of test sample	Test sample 1	Test sample 2	Test sample 3	Mean value f_{cm}
Compression strength f_{ci} [N/mm^2]	31.80	29.60	31.70	31.03

Table 1.
Compressive strength testing results for concrete.

Reinforcing steel (f_{yk} = 500 MPa) was procured locally in the market. As part of this research, the mechanical characteristics of the applied reinforcing steel were tested. From the results, it can be seen that the mean yield stress value of the steel is without large deviations and is 660.33 MPa and is significantly above f_{yk} = 500 MPa. On the other hand, EN 1992-1-1 [13] in Section 3.2.7 (4) specifies that the calculation value of the modulus of elasticity of reinforcing steel of E_s = 200 GPa can be assumed. By analyzing the results of the reinforcing steel test, it was established that the mean value of the modulus of elasticity of the applied reinforcing steel is significantly lower and amounts to 127.33 GPa, with minimal deviations from the mean value (125 GPa minimum value and 130 GPa - maximum value), which is significantly lower than the assumption in EN 1992-1-1 [13].

3.4 Production of test samples

Figures 4 and **5** show the phases of production of RC beam samples. The remaining two series of beams are used for other studies which are not part of this research.

Processed timber was used to produce formwork. The production of test beams was conducted at the local concrete plant by licensed staff. The conditioning of test beams was conducted for 28 days of age, according to the standard specifications for such testing.

Figure 4.
Phases of production of test samples.

DOI: http://dx.doi.org/10.5772/intechopen.1002342

Figure 5.
Produced test samples of beams.

3.5 Testing equipment

The following equipment shown in **Figure 6** was used during the test:

- Hydraulic gantry press;
- Digital flexogauge with a resolution of 0.001 mm;
- Optical flexogauge with a resolution of 0.01 mm;
- Optical magnifier - magnification 200 times.

The test equipment, including measuring devices, was calibrated prior to testing, so that the data would be credible.

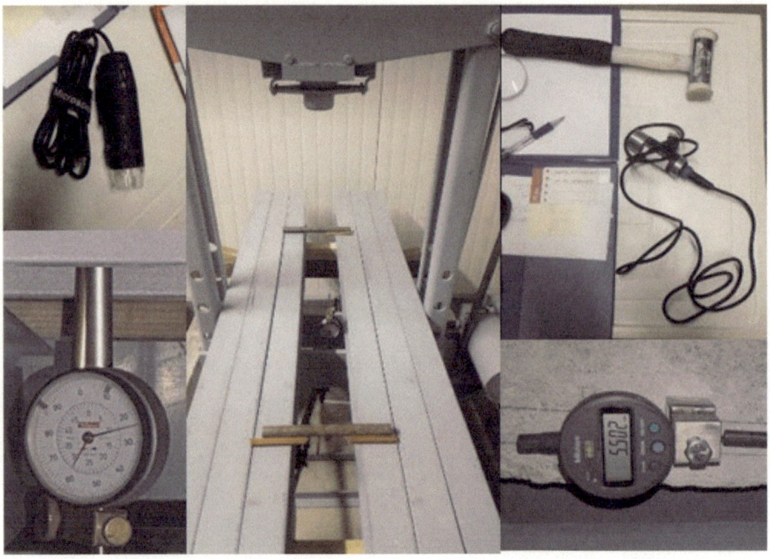

Figure 6.
Laboratory equipment.

4. Conducting experiment

The simulation of the loading of the samples by hydraulic gantry press was done on statical system of a simple beam, with a span of 1.30 m, with concentrated forces in a third of the span according to **Figure 7**.

The load on the beam is applied gradually in increments according to **Table 2** up to the sample failure and then the test samples would be unloaded.

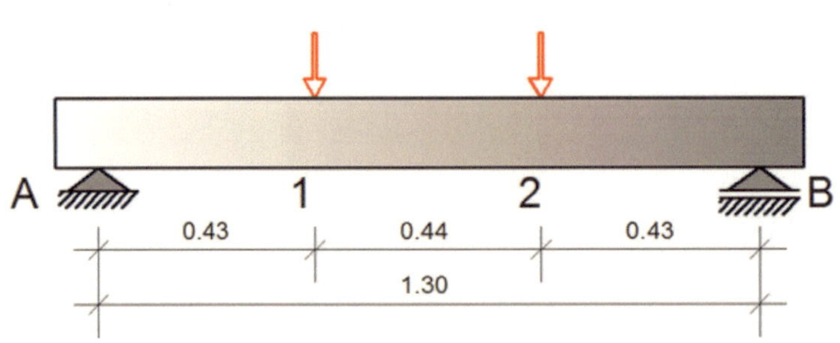

Figure 7.
Statical system (dimensions in m).

Phase	Hydraulic gantry press stress [MPa]	Hydraulic gantry press force [kN]	Maximal shear force Vmax [kN]	Maximal bending moment Mmax [kNm]
0	0.00	0.00	0.00	0.00
1	1.25	2.97	1.48	0.64
2	2.50	5.94	2.97	1.29
3	3.75	8.90	4.45	1.93
4	5.00	11.87	5.94	2.57

Phase	Hydraulic gantry press stress [MPa]	Hydraulic gantry press force [kN]	Maximal shear force Vmax [kN]	Maximal bending moment Mmax [kNm]
5	6.00	14.25	7.12	3.09
6	7.00	16.62	8.31	3.60
7	8.00	19.00	9.50	4.12
8	9.00	21.37	10.69	4.63
9	10.00	23.75	11.87	5.15
10	11.00	26.12	13.06	5.66
11	12.00	28.50	14.25	6.17
12	13.00	30.87	15.44	6.69
13	14.00	33.24	16.62	7.20
14	15.00	35.62	17.81	7.72
15	16.00	37.99	18.90	8.23
16	17.00	40.37	20.18	8.75
17	18.00	42.74	21.37	9.26
18	19.00	45.12	22.56	9.78
19	20.00	47.49	23.75	10.29

Table 2.
Gradual loading plan.

Since the beam has its self-weight, as well as the weight of the apparatus, this constant load cause specific effects, so they also need to be taken into account. The statical system with the self-weight of the beam is shown in **Figure 8** and apparatus for load transfer in **Figure 9**.

Apparatus for load transfer – concentric force G = *0.11 kN.*

Self-weight distributed load *g*:

$$g = 0.10 \times 0.20 \times 1.4 \times 25/1.3 = 0.54\ \mathrm{kN/m'} \tag{1}$$

Maximal shear force of self-weight and apparatus $V_{g,max}$:

$$V_{g,\,max} = 0.54 \times 1.3/2 + 0.11 = 0.46\ \mathrm{kN} \tag{2}$$

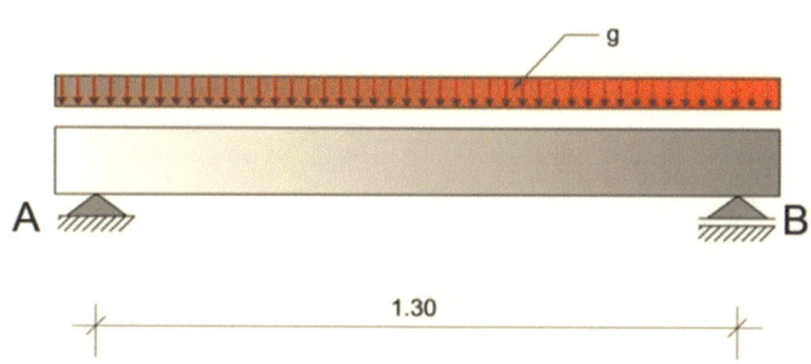

Figure 8.
Self-weight of the beam (dimensions in m).

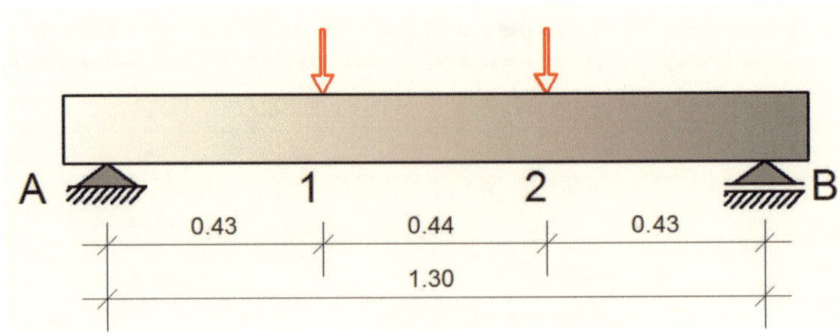

Figure 9.
Apparatus for load transfer (dimensions in m).

Maximal bending moment of self-weight and apparatus $M_{g,max}$:

$$M_{g,\,max} = 0.54 \times 1.3^2/8 + 0.11 \times 1.3/3 = 0.16\ \text{kNm} \tag{3}$$

During the test, and for each load increment of the test beams, the deflection is measured with an optical deflection meter with a resolution of 0.01 mm (**Figure 10**).

Figure 10.
Optical deflection meter during testing measuring deflection.

Figure 11.
Digital flexometer meter during testing measuring the axial deformation at the position of the longitudinal reinforcement.

During the test, the axial deformation at the position of the main longitudinal tensile reinforcement is also measured with a digital flexometer with a resolution of 0.001 mm (**Figure 11**).

Experienced and licensed engineers and laboratory technicians conducted quality control during the tests.

5. Experimental results

During the tests in the laboratory, three beams were tested, which were marked as B1, B2, and B3. The test results are presented in the following tables and diagrams.

5.1 Test sample B1

Test sample B1 proved to be the weakest sample in the series. By increasing the bending moment over 3.76 kNm, the first visible cracks appear on the supports. Oblique cracks are also visible, which predominate near the support, which shows that concrete as a material has exhausted its shear capacity. Further increase of the bending moment leads to their propagation and expansion. The loss of flexural bearing capacity of the Test Sample B1 occurs at a bending moment of 5.31 kNm. Namely, then the strain of the reinforcement is significantly over 2‰ (**Figure 13**), which means that the stress in the

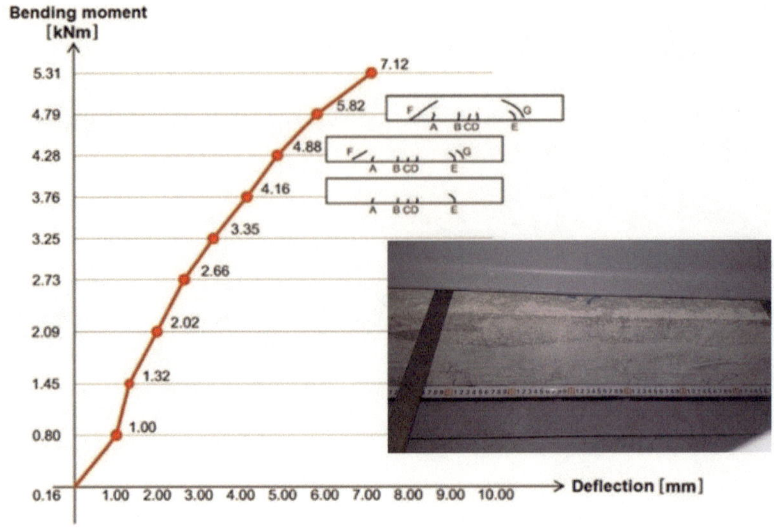

Figure 12.
Diagram of deflections for test sample B1.

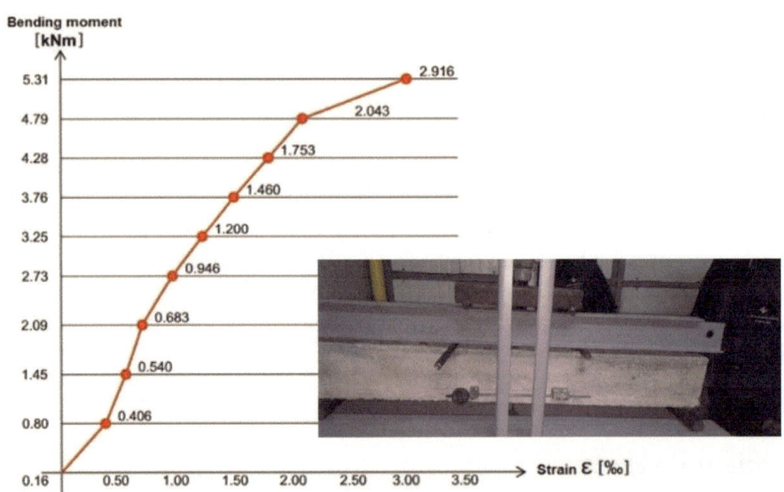

Figure 13.
Diagram of strains for test sample B1.

reinforcement has reached the tensile yield limit, and after that the reinforcement strain suddenly increases. After unloading, the residual deflection is 3.06 mm (**Figures 12** and **13**), and the plastic strain of the reinforcement is 0.353‰ (**Table 3**).

5.2 Test sample B2

Test Sample B2 shows more favorable behavior in terms of the appearance of the first cracks, which occur at a bending moment of 4.28 kNm, when the first cracks A,

Test sample B1					
Phase	**Shear force**	**Bending moment**	**Deflection**	**Axial deformation**	**Strain ε**
	[kN]	[kNm]	[mm]	[mm]	‰
0	0.46	0.16	0.00	0.000	0.000
1	1.94	0.80	1.00	0.122	0.406
2	3.43	1.45	1.32	0.162	0.540
3	4.91	2.09	2.02	0.205	0.683
4	6.40	2.73	2.66	0.284	0.946
5	8.04	3.25	3.35	0.360	1.200
6	8.77	3.76	4.16	0.438	1.460
7	9.96	4.28	4.88	0.526	1.753
8	11.15	4.79	5.82	0.613	2.043
9	12.33	5.31	7.12	0.875	2.916
Load relief	0.46	0.16	3.06	0.106	0.353

Table 3.
The test results for test sample B1.

B, C, and D appear (**Figure 14**). By increasing the bending moment to 4.69 kNm, oblique cracks F, G appear and H (**Figure 14**), which show that the concrete has lost its shear capacity, and their further propagation creates concrete struts. Flexural failure of the beam occurs at a bending moment of 5.31 kNm. After unloading, the residual deflection is 3.35 mm (**Figure 15** and **Table 4**).

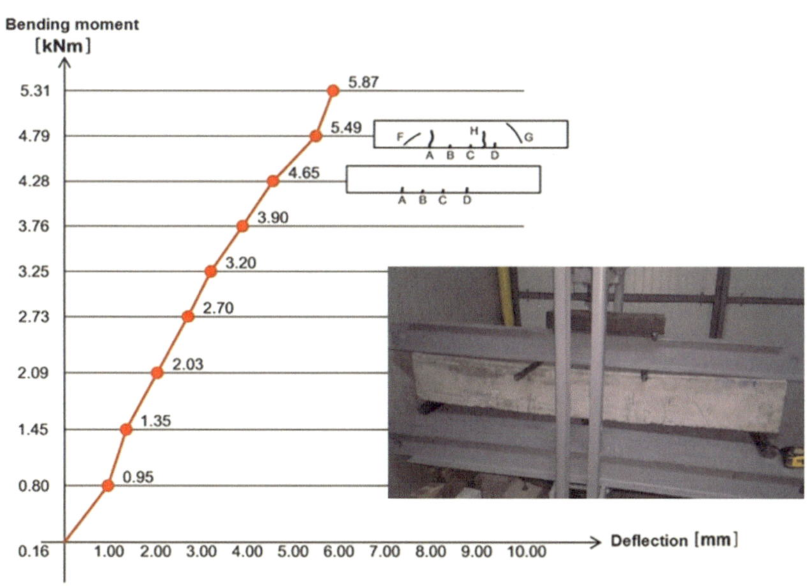

Figure 14.
Diagram of deflections for test sample B2.

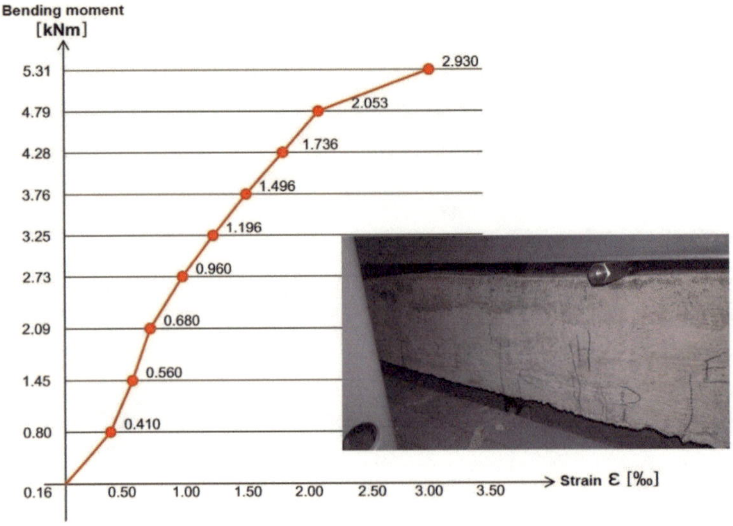

Figure 15.
Diagram of strains for test sample B2.

Test sample B2					
Phase	Shear force	Bending moment	Deflection	Axial deformation	Strain ε
	[kN]	[kNm]	[mm]	[mm]	‰
0	0.46	0.16	0.00	0.000	0.000
1	1.94	0.80	0.95	0.123	0.410
2	3.43	1.45	1.35	0.168	0.560
3	4.91	2.09	2.03	0.204	0.680
4	6.40	2.73	2.70	0.288	0.960
5	8.04	3.25	3.20	0.359	1.196
6	8.77	3.76	3.90	0.449	1.496
7	9.96	4.28	4.65	0.521	1.736
8	11.15	4.79	5.49	0.616	2.053
9	12.33	5.31	5.87	0.879	2.930
Load relief	0.46	0.16	3.35	0.140	0.466

Table 4.
The test results for test sample B2.

5.3 Test sample B3

Test Sample B3 showed similar results to beam B2. The results of the experimental analysis of Test Sample B3 are presented in tabular (**Table 5**) and graphical form (**Figures 16** and **17**). The failure of the Test Sample B3 occurs when a bending moment of 5.31 kNm is applied. After unloading, the residual deflection is 3.16 mm,

Test sample B3					
Phase	**Shear force**	**Bending moment**	**Deflection**	**Axial deformation**	**Strain ε**
	[kN]	[kNm]	[mm]	[mm]	‰
0	0.46	0.16	0.00	0.000	0.000
1	1.94	0.80	0.32	0.117	0.390
2	3.43	1.45	0.73	0.158	0.526
3	4.91	2.09	1.14	0.213	0.710
4	6.40	2.73	2.73	0.303	1.010
5	8.04	3.25	3.38	0.363	1.210
6	8.77	3.76	4.10	0.453	1.510
7	9.96	4.28	4.88	0.483	1.610
8	11.15	4.79	5.68	0.613	2.043
9	12.33	5.31	6.41	0.873	2.910
Load relief	0.46	0.16	3.16	0.134	0.446

Table 5.
The test results for test sample B3.

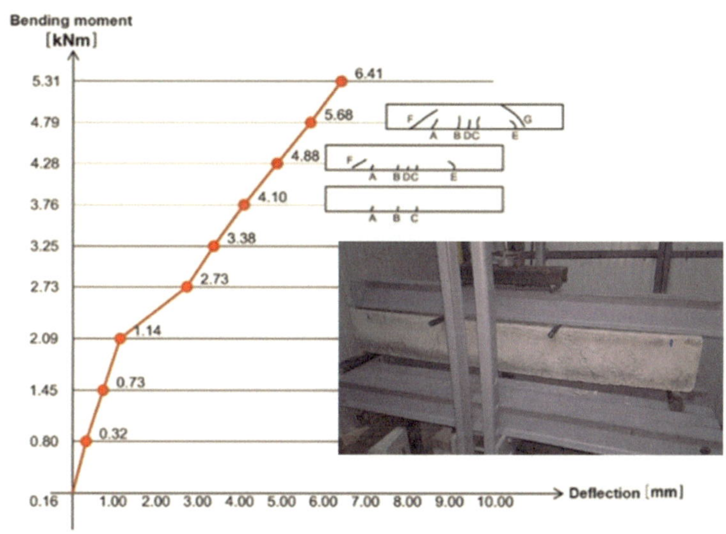

Figure 16.
Diagram of deflections for test sample B3.

and the plastic strain of the reinforcement is 0.446‰. Based on the obtained experimental results, an experimental flexural capacity of all test samples is 5.31 kNm.

5.4 Width of cracks

One of the basic assumptions in the design of reinforced concrete structures in flexure is that the concrete does not take over the tensile forces and that all tensile

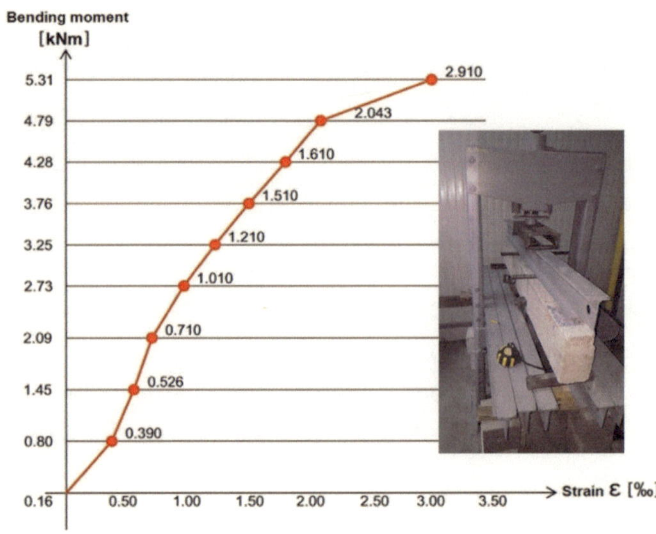

Figure 17.
Diagram of strains for test sample B3.

stresses are taken over by the steel reinforcement. Cracks occur when actual tensile stresses exceed the tensile strength of concrete. In general, cracks in the tension zone of reinforced concrete do not affect the flexural capacity of the structure to a large extent. On the other hand, when designing reinforced concrete structures to serviceability limit states (SLS), the width of cracks is also considered, especially in facilities where width of cracks affects the corrosion of reinforcement, external appearance, and impermeability to liquids or gases. The limit width of cracks for reinforced concrete structures ranges from 0.3 to 0.4 mm, depending on the exposure class of the concrete. In this research, during the experimental analysis, crack propagation was also observed, considering that it was feasible, although it was not the primary goal of the research.

The widths of the cracks in the concrete were observed using an optical magnifier with a magnification of 200 times. The following figure shows the measured crack widths at a bending moment of 3.76 kNm.

Figure 18 shows that the width of the cracks is up to 0.30 mm at a bending moment of 3.76 kNm. However, by increasing the bending moment to 4.28 kNm, the cracks widen from 0.41 to 0.85 mm, which is shown in **Figure 19**.

6. Determination of flexural capacity of beams with modulus of elasticity of steel reinforcement E_s = 200 GPa

The determination of the flexural capacity of the beams is conducted using a rectangular stress distribution in the concrete section, according to EN 1992-1-1, Section 3.1.7 (13) (**Figure 20**).

The determination of flexural capacity assumes a system in equilibrium. In this case, the compressive force in the concrete and the tensile force in reinforcement are equal.

DOI: http://dx.doi.org/10.5772/intechopen.1002342

Figure 18.
Measured crack widths at a bending moment of 3.76 kNm (in mm).

Figure 19.
Measured crack widths at a bending moment of 4.28 kNm (in mm).

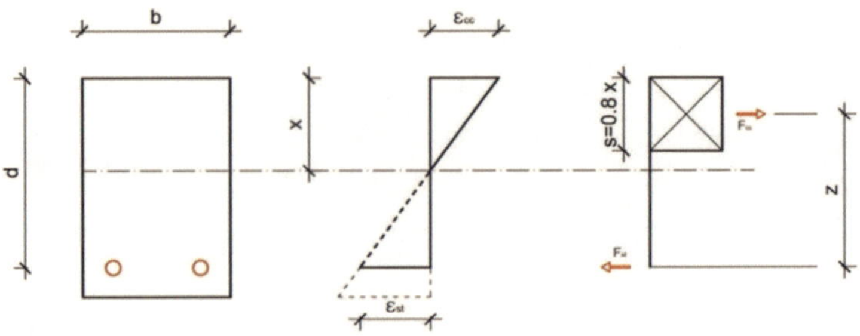

Figure 20.
Rectangular stress distribution in the concrete section [16].

$$\boldsymbol{F}_{cc} = \boldsymbol{F}_{St} \tag{4}$$

Then, the compressive force in concrete is:

$$\boldsymbol{F}_{cc} = \boldsymbol{\sigma}_c \cdot \boldsymbol{A} \tag{5}$$

where
σ_c - compression stress in concrete;
A - concrete area in compression.
Compression stress in concrete is:

$$\sigma_c = \frac{\alpha \cdot f_{ck}}{\gamma_c} = \frac{0.85 \cdot f_{ck}}{1.5} \tag{6}$$

while concrete area in compression is:

$$A = b \cdot s \tag{7}$$

and tensile force is calculated as:

$$F_{St} = \sigma_s \cdot A_s \tag{8}$$

where:
σ_s - tensile stress in reinforcement.
A_s - cross-sectional area of reinforcing steel.
Tensile stress in reinforcement is:

$$\sigma = \frac{f_{yk}}{\gamma_s} = \frac{f_{yk}}{1.15} \tag{9}$$

while flexural capacity is:

$$M_{Rd} = F_{st} \cdot z \tag{10}$$

In this specific case of the cross section of a reinforced concrete beam with dimensions of width 100 mm and height 200 mm, made of concrete class C 25/30, reinforced with longitudinal reinforcement in both zones 2 Ø 8, of reinforcing steel f_{yk} = 500 MPa, the calculation of the flexural capacity is as follows:

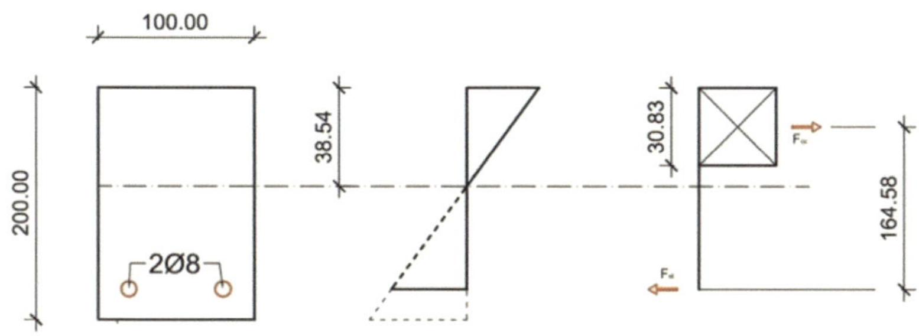

Figure 21.
Rectangular stress distribution and dimensions in actual concrete section (dimensions in mm).

$$\boldsymbol{F_{cc}} = \boldsymbol{F_{St}} \tag{11}$$

$$\frac{0.85 \cdot f_{ck}}{1.5} \cdot b \cdot s = \frac{f_{yk}}{1.15} \cdot A_s \tag{12}$$

$$0.567 f_{ck} \cdot b \cdot s = 0.87 f_{yk} \cdot A_s \tag{13}$$

Based on that, the height of the compressive concrete block is:

$$s = \frac{0.87 f_{yk} \cdot A_s}{0.567 f_{ck} \cdot b} = \frac{0.87 \cdot 500 \cdot 100.48}{0.567 \cdot 25 \cdot 100} = 30.83\ mm \tag{14}$$

The arm of internal forces z is obtained as:

$$z = d - 0.5 \cdot s = 180 - 0.5 \cdot 30.83 = 164.58\ mm \tag{15}$$

Based on the obtained values, the flexural capacity of the cross section is (**Figure 21**):

$$M_{Rd} = F_{St} \cdot z = 0.87 f_{yk} \cdot A_s \cdot z = 0.87 \cdot 500 \cdot 100.48 \cdot 164.58 = 7.19\ kNm \tag{16}$$

7. Model for determining the flexural capacity of RC beams reinforced by steel with reduced modulus of elasticity

Table 6 displays an overview of the experimentally recorded axial deformations of the test beams with the mean value of strains depending on the phase of loading of test beams to the failure.

Data on strains are also shown in the diagram in **Figure 22**, which shows a significant concurrence of experimental strains for all three samples of reinforced concrete beams.

Considering the mean strain was 2919‰ at the moment of failure of test beams, and the mean value of the modulus of elasticity of the reinforcing steel with reduced modulus of elasticity (127.33 GPa), the tensile stress in the reinforcing steel was:

$$\sigma_s = E_{s(mean)} \cdot \epsilon_{mean} = 127.33 \cdot 2.919 \cdot 10^{-3} = 371.68\ MPa \tag{17}$$

Phase	Shear force [kN]	Bending moment [kNm]	Axial deformation			Mean strain ε ‰
			Test Sample B1 [mm]	Test Sample B2 [mm]	Test Sample B3 [mm]	
0	0.46	0.16	0.000	0.000	0.000	0.000
1	1.94	0.80	0.406	0.410	0.390	0.402
2	3.43	1.45	0.540	0.560	0.526	0.542
3	4.91	2.09	0.683	0.680	0.710	0.691
4	6.40	2.73	0.946	0.960	1.010	0.972
5	8.04	3.25	1.200	1.196	1.210	1.202
6	8.77	3.76	1.460	1.496	1.510	1.489
7	9.96	4.28	1.753	1.736	1.610	1.700
8	11.15	4.79	2.043	2.053	2.043	2.047
9	12.33	5.31	2.916	2.930	2.910	2.919

Table 6.
Comparison of experimental test results for axial deformation.

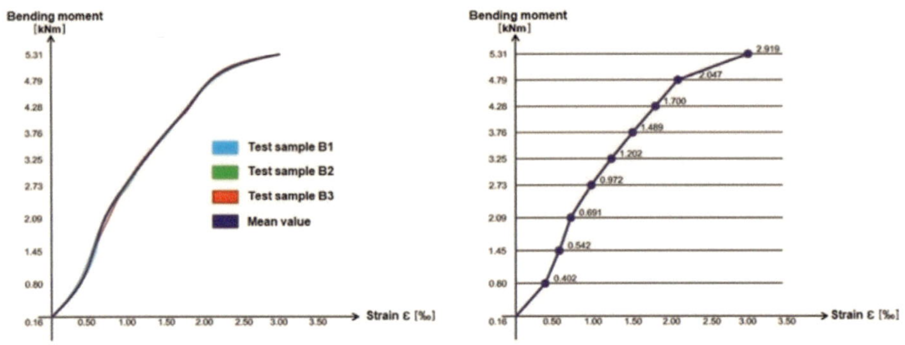

Figure 22.
Concurrence on experimental strains for all three samples of test reinforced concrete beams.

Furthermore, this means that the tensile force in the steel reinforcement is:

$$F_{st} = \sigma_s \cdot A_s = 371.68 \cdot 100.48 \cdot 10^{-3} = 37.346\, kN \tag{18}$$

and based on the equilibrium of forces in the cross section

$$F_{cc} = F_{St} \tag{19}$$

that is

$$0.567 f_{ck} \cdot b \cdot s = 37.346\, kN \tag{20}$$

so, the depth of the compression stress block of concrete s is

$$s = \frac{F_{st}}{0.567 f_{ck} \cdot b} = \frac{37.346}{0.567 \cdot 25 \cdot 100} = 26.35\, mm \tag{21}$$

and the arm of internal forces is equal to:

$$z = d - 0.5 \cdot s = 180 - 0.5 \cdot 26.35 = 166.82\, mm \tag{22}$$

Based on this, the compression stress in concrete is:

$$\sigma_c = \frac{F_{St}}{0.567 \cdot b \cdot s} = \frac{37.346 \cdot 10^3}{0.567 \cdot 100 \cdot 26.35} = 25.00\, MPa = f_{ck} = 25.00\, MPa \tag{23}$$

which exhausted the compression capacity of the concrete, so the compression stress block of the concrete fails by brittle fracture through concrete, and confirmed by experiment, so there is a complete concurrence of the experimental and calculated data on the flexural bearing capacity of the section reinforced by steel with reduced modulus of elasticity.

8. Conclusion

The experimental research showed that test RC beams reinforced by steel with reduced modulus of elasticity (127,33 GPa) have reduced flexural capacity in comparison with designed flexural capacity of beams reinforced by steel with "normal" modulus of elasticity (200 GPa) by 26.15% (**Table 7**).

The experiment showed that reduced modulus of elasticity of steel in reinforced concrete beams leads to faster increase in the strain of the reinforcement, which additionally leads to the displacement of the neutral axis toward the extreme compression fiber of the section, reducing the compression stress block of the concrete, and as a result of the equilibrium of forces in the cross section, consequently leads to the brittle failure of the concrete, which reduces the flexural capacity of the cross section in comparison to the designed flexural capacity according to EN 1992-1-1. This is especially pronounced in the case of reinforced concrete beams with smaller dimensions of cross section.

The results of this research can be applied if reinforcing steel with a reduced modulus of elasticity has already been embedded into the reinforced concrete beams, all with the aim of determining the flexural capacity of the already constructed beam, as the basis for concluding whether it is necessary to repair or rertrofit the concrete beam or even the entire structure.

The calculation model proposed in this research can also be used for further research for beams with larger cross-sectional dimensions and a larger amount and diameter of embedded steel reinforcement with a reduced modulus of elasticity, as well as for RC slabs in order to confirm or modify the model.

The reasons for the presence of steel reinforcement on the local markets with a reduced modulus of elasticity are not discussed here. However, the installation of this

Experimental flexural capacity for test beams with reduced modulus of elasticity [kNm]	Designed flexural capacity for test beams according to EN 1992-1-1 [kNm]	Percentage [%]
5.31	7.19	73.85

Table 7.
Comparison of experimental flexural capacity test results for test beams with reduced modulus of elasticity and designed flexural capacity for test beams according to EN 1992-1-1.

type of steel reinforcement in concrete elements without prior insight into the mechanical properties of the steel reinforcement before installation can lead to serious consequences in the service of the building or facility. In this sense, in areas and countries with inadequate control of the market for construction materials, it is necessary to examine the mechanical properties of the steel reinforcement before installation and make sure that it corresponds to the mechanical properties required by the design. In this sense, when it comes to steel reinforcement, it is not enough to check the yield strength of the steel reinforcement only, but also other specified mechanical properties, especially the modulus of elasticity, which this research has shown.

Author details

Sanin Dzidic* and Aldin Mahmutovic
Faculty of Technical Sciences, Department of Civil Engineering, University of Bihac, Bihac, Bosnia and Herzegovina

*Address all correspondence to: ninsa_d@hotmail.com

References

[1] Dzidic S, Mahmutovic A. Analysis of the possibilities for retrofit of RC beams by traditional and modern methods. In: Proceedings of the 18th International Symposium of Macedonian Association of Structural Engineers MASE 18. Ohrid, North Macedonia: Macedonian Association of Structural Engineers; 2019. pp. 270-279. ISBN 978-608-4510-36-9

[2] Mahendran M. The modulus of elasticity of steel - Is it 200 GPa? In: Thirteenth International Specialty Conference on Cold-Formed Steel Structures. St. Louis, Missouri, USA: Missouri University of Science and Technology; 1996

[3] Miljojkovic J et al. Determining elastic modulus of the material by measuring the deflection of the beam loaded in bending. Technical Gazette 24. 2017;**4**: 1227-1234. DOI: 10.17559/TV-20170609133537

[4] Radovic E, Lara-Curzio E, Riester L. Comparison of different experimental techniques for determination of elastic properties of solids. Materials Science and Engineering. 2004;**368**(1-2):56-70. DOI: 10.1016/j.msea.2003.09.080

[5] Bamber MJ, Cooke KE, Mann AB, Derby B. Accurate determination of Young's modulus and Poisson's ratio of thin films by a combination of acoustic microscopy and nanoindentation. Thin Solid Films;**398-399**:299-305. DOI: 10.1016/S0040-6090(01)01341-4

[6] Rho JY, Ashman RB, Charles HT. Young's modulus of trabecular and cortical bone material: Ultrasonic and microtensile measurements. Journal of Biomechanics. 1993;**26**(2):111-119. DOI: 10.1016/0021-9290(93)90042-D

[7] Zhong C, Umesh G, Jinwoo L, Wagoner RH. Variation and consistency of Young's modulus in steel. Journal of Materials Processing Technology. 2016; **227**:227-243. DOI: 10.1016/j.jmatprotec.2015.08.024

[8] Lems W. The Change of Young's Modulus after Deformation at Low Temperature and its Recovery", Ph.D Dissertation. Delf, Netherlands: Delft University of Technology; 1963

[9] Morestin F, Boivin M. On the necessity of taking into account the variation in the young modulus with plastic strain in elastic-plastic software. Nuclear Engineering and Design. 1996; **162**(1):107-116. DOI: 10.1016/0029-5493(95)01123-4

[10] Yoshida F, Uemori T, Fujiwara K. Elastic-plastic behavior of steel sheets under in-plane cyclic tension–compression at large strain. International Journal of Plasticity. 2002;**18**(5–6): 633-659. DOI: 10.1016/S0749-6419(01)00049-3

[11] Cleveland RM, Ghosh AK. Inelastic effects on springback in metals. International Journal of Plasticity. 2002; **18**(5–6):769-785. DOI: 10.1016/S0749-6419(01)00054-7

[12] Shayan JA, Peiris R, Coorey RV. Study of elastic properties of reinforcing steel bars. In: Proceedings of the Technical Sessions. Vol. 26. Sri Lanka: Institute of Physics; 2010. pp. 67-74

[13] EN 1992-1-1:2004 Eurocode 2. Design of concrete structures. In: Part 1-1: General Rules – Structural Fire Design and Rules for Buildings. Avenue Marnix 17, B-1000 Brussels, Belgium: CEN European Committee for Standardization, Management Centre; 2004

[14] EN 1992-1-2:2004. Eurocode 2, Design of concrete structures. In: Part 1-2: General Rules – Structural Fire Design. rue de Stassart, 36 B 1050 Brussels, Belgium: CEN European Committee for Standardization, Management Centre: Management Centre; 2004

[15] EN 206:2013. Concrete-specification, performance, production and conformity. In: CEN European Committee for Standardization. Avenue Marnix 17, B-1000 Brussels, Belgium: CEN-CENELEC Management Centre; 2013

[16] Mosley B, Bungey J, Hulse R. Reinforced concrete design to Eurocode 2 - seventh edition. In: Palgrave Macmillan in the UK. Hampshire RG21 6XS, UK: Palgrave Mac-Millan; 2012. ISBN-13: 978–0–230–30285–3

Chapter 3

Fracture Behavior of Structural Steels under Earthquake Dynamic Loadings

Alireza Khalifeh, Habib Danesh Manesh and Abolghasem Dehghan

Abstract

The satisfactory design of the components is highly dependent on the adequate knowledge of the material behavior and operational conditions. For the structures under earthquakes, often this information is not available, is incomplete or inaccurate, and leads to increases the risk of the possible failures. The extensive brittle fracture of steel structures during the Northridge earthquake (USA, 1994) and Kobe earthquake (Japan, 1995) highlighted many of these deficiencies. The investigations have shown that the failures were caused by combination effects of high strain rate, welding defects, welding residual stress, and seismic loads. In this chapter, the effects of strain rates on mechanical properties of steel materials have been discussed. Welding defects act as cracks and cause the structures to fail at loads lower than design stress. Thus, the issue has been evaluated from the view point of failure mechanics. Welding processes produced residual stresses in the weldments. These regions have higher stresses triaxiality and will be prone to brittle fracture. Therefore, the role of residual stresses in the failure of steel structures is well expressed. The contents given in this chapter can be of great help in preventing the failure of structures during an earthquake and the occurrence of human and financial losses.

Keywords: structural steels, dynamic loadings, mechanical properties, fracture behavior, residual stress

1. Introduction

An earthquake is an unexpected, unpredictable and uncontrollable natural event that occurs in earthquake zones. High loading rate, load application rate 0.1–1/s, changes the mechanical behavior of structural steels. This makes the used metal structures susceptible to brittle fracture and causes many human and financial losses [1–7].

If a structure can withstand large inelastic deformation change, without instability, it is said that it has shown ductile behavior. Design engineers believe that by choosing materials and connections with high flexibility and proper strength, fracture caused by earthquakes can be prevented. Many buildings have been designed and built based on this vision. But extensive brittle fracture during two earthquakes in 1994 in Northridge, America and 1995 in Kobe, Japan cause doubt on this idea [8, 9]. Many cases of such unpredictable brittle fracture have been reported in the lower-flange welds of steel moment resistant frames (WSMFs) as it is seen in **Figures 1** and **2** [12, 13]. WSMFs are often constructed using large rolled sections of A36 (beams) and A572 (columns) steel with flux-core field welds (E70T-4) utilized for transferring the primary load [14]. This prompted the idea of new research programs to study the causes of these fractures and to propose needed changes in design and construction procedures [12, 15, 16].

Comprehensive and extensive field and laboratory studies have been conducted regarding the causes of fracture of structures during the two events of Northridge and Kobe earthquakes. The results of the investigations showed that the unexpected and brittle failures in the steel structures during earthquake dynamic loadings were caused by factors such as high strain rate, welding defects, improper connection design, low toughness of the materials used, and seismic loads [6, 16–21]. The residual stress

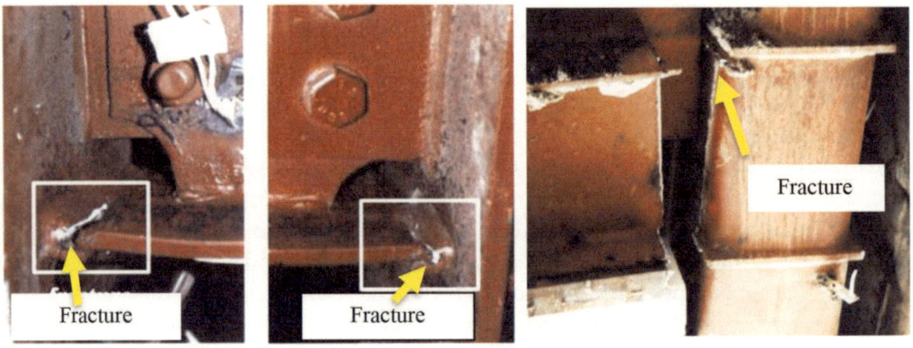

Figure 1.
Example of steel structural failure due to earthquakes [10].

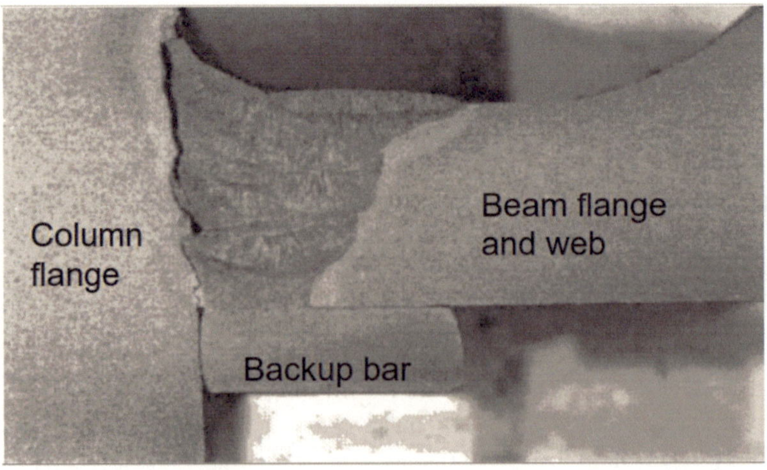

Figure 2.
Photograph of a typical brittle fracture in steel structures [11].

caused by welding is another important factor involved in fracture of structural steels under earthquake type dynamic loadings [22–25]. The extent of residual stresses can sometimes reach the yield stress of the material [26]. Therefore, the effect of the presence of residual stresses should also be considered in the fracture studies of steel structures.

In this chapter, the earthquake type dynamic loading is first introduced. Then, the effects of high strain rate on mechanical properties of structural steels are expressed. In next step, the effects of welding defects on fracture behavior of steel structures are discussed using fracture mechanic criteria's. Finally, the residual stresses and their effects are explained in detail.

2. Strain rate effects on mechanical behavior

Many studies have been done to determine the plasticity characteristics metals at high strain rates. This information is vital in the performance of structures in earthquake and impact conditions [27, 28]. The strain rate, $\dot{\varepsilon}$ is defined as follows:

$$\dot{\varepsilon} = \frac{d\varepsilon}{dt} \tag{1}$$

The conventional unit of strain rate is s^{-1} [29].

Increasing the strain rate causes the yield point to be observed in steels that do not have a yield point in the static state. For most materials, increasing the strain rate causes an increase in flow stress. The amount of increase depends on the type of material and temperature. For most materials, the effect of strain rate on the flow stress curve, σ, at a constant strain and temperature can be described by a power expression [27, 29]:

$$\sigma = C\dot{\varepsilon}^m \tag{2}$$

where σ is stress, $\dot{\varepsilon}$ is strain rate, and m is called strain rate sensitivity.

Kaufman and Chang have estimated that in the absence of sources of stress concentration, the value of the nominal strain rate ($\dot{\varepsilon}$) for welded joints during an earthquake is in the range of 0.01 to 1/s [30, 31]. This amount of strain rate is 1000 to 10,000 times higher than the quasi-static laboratory conditions. Meanwhile, the presence of stress concentration in front of a crack in the weld of the structure can increase the strain rate to 0.05 to 1/s on the local volume of the material. This amount of strain rate increases the flow stress and reduces the fracture toughness of steel structures and welds and facilitates the conditions of brittle failure (**Figure 3**) in welded joints [32].

In general, the yield stress of structural steels increases with increasing the strain rate and decreasing temperature [27]. The study of the effect of strain rate on the flow properties of three main types of materials used in structures (A36 beams, A572 columns and E70T-4 welds) was done by Kaufman, Fisher Rolf and Barson (**Figure 4**) [21]. Accordingly, the lower strength A36 material exhibits the most strain sensitivity compares with higher strength steels grades such as A572 steel and E70T-4. The results shown in **Figure 4** are based on measured σ_{ys}^{static} by Kaufman and Fisher and the empirical model (Eq. (3)) provided by Rolf and Barson [11].

Figure 3.
Typical photograph of a typical brittle fracture in steel structures.

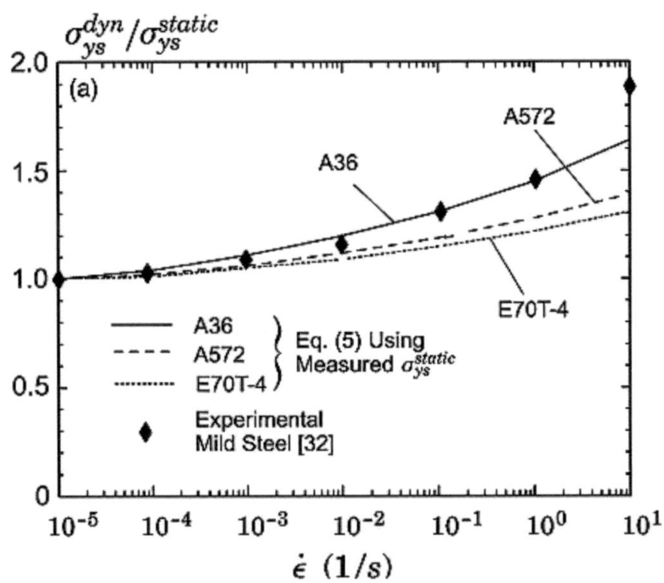

Figure 4.
Effects of strain rate on the yield stress of structural steel connections [32, 33].

$$\frac{\sigma_{ys}^{dyn}}{\sigma_{ys}^{static}} = 1 + \left[\frac{\frac{2}{3} \times 10^6}{\sigma_{ys}^{static} \, log \left\{ 2 + \frac{2}{3} \times 10^6 \frac{\sigma_{ys}^{static}}{\dot{\varepsilon}} \right\} (T + 273)} - \frac{\sigma_{ys}^{dyn}}{\sigma_{ys}^{static}} \right] \tag{3}$$

where σ_{ys}^{dyn} is the static yield strength at room temperature (MPa), T (°C) and $\dot{\varepsilon}$ refers to the temperature and strain rate of interest, respectively.

According to **Figure 4**, For A36 steel the yield strength increases by 45% at the strain rate of 1/s and by 65% at the strain rate of 10/s. For A572 steel, a 40% increase in yield strength has been observed for a strain rate of 10/s. This sensitivity value for the weld metal obtained from E-70 T-4 with higher strength than A36 and A572 reaches 25% at a strain rate of 10 seconds, which is a much lower value [32].

2.1 Mechanism

The increase in yield strength of steels due to the increase of strain rate can be described using dislocations theories. The equation for the movement dependence of dislocations on the strain rate is considered as follows [27, 34].

$$\dot{\varepsilon} = \rho b \overline{v} \tag{4}$$

where ρ is the density of dislocations, $\overline{v}$ is the average velocity of dislocations and b is the Burgers vector. On the other hand:

$$\overline{v} \propto \tau^{m} \tag{5}$$

where τ is the shear stress and m is a constant whose value is equal to 35 for iron [27, 34]. In low carbon steels, the dislocations are locked with carbon and nitrogen in the structure. The only way for $\rho b \overline{v}$ to adapt to the imposed strain rate is to increase $\overline{v}$. According to Eq. (5), this state requires applying high stress which lead to increasing the yield stress [27, 34].

Research has shown that the increasing strain rate also affects the fracture toughness of materials [35, 36]. The graph of the effect of strain rate on the fracture toughness of a steel is given in **Figure 5**. As it can be seen in the picture, with the increase in loading rate, the temperature of the transition from ductile to brittle of steels has shifted to higher values. This means that the brittle failure of the material under dynamic loads occurs at a higher temperature [35].

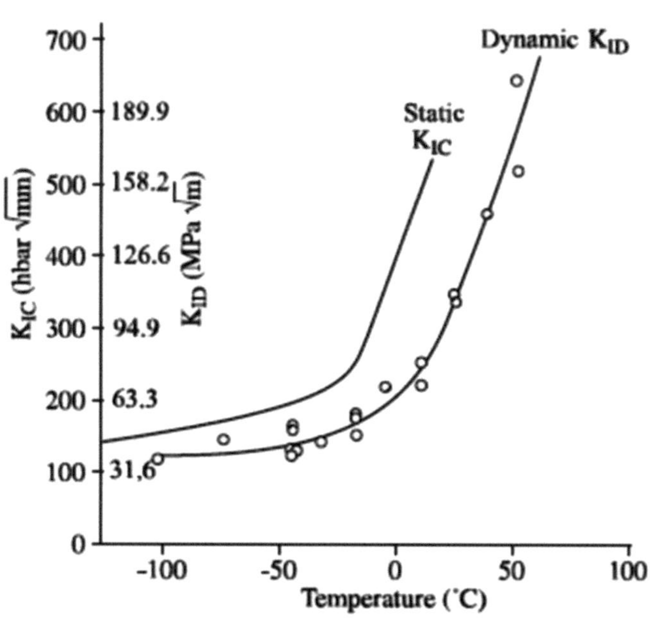

Figure 5.
The effect of loading rate and temperature on fracture toughness of a carbon steel material [35].

During an earthquake fire may also occur. In a fire, the high thermal conductivity of steel can spread the heat in the steel structure from a local part to the whole structure in a short time. The mechanical properties of steel structure members will be affected by high temperature impressively. This may have caused numerous economic losses and casualties. Yet, when the fire is extinguished, some fire-affected structural steels may still be used, and what determines they are reusable or not is whether their residual mechanical properties meet the engineering requirements. Therefore, it is of great importance to study the post-fire mechanical properties of structural steels [37, 38]. An equation is proposed by Johnson-Cook, which well express the simultaneous effect of temperature and strain rate on flow stress behavior can be used [39, 40]:

$$\sigma = [A + B\varepsilon^n][1 + Cln\dot{\varepsilon}^*][1 - T^{*m}] \quad (6)$$

where σ is the stress, ε is the plastic strain, m is the sensitivity coefficient to the strain rate, A, B, C are the constants of the material. The values of $\dot{\varepsilon}^*$ and T^{*m} are expressed as follows:

$$\dot{\varepsilon}^* = \dot{\varepsilon}/\dot{\varepsilon}_0 \quad \text{and} \quad T^* = \frac{(T - T_0)}{(T_m - T_0)} \quad (7)$$

where T_m is the melting point of the material, T_0 is the ambient temperature, $\dot{\varepsilon}_0$ is the static strain rate and $\dot{\varepsilon}^*$ is the dynamic strain rate.

3. Effects of welding defects and fracture mechanic

Welding defects are among the important factors of structural failure under earthquake loads [17, 31, 41, 42]. The most important welding defect of welded joints is the incomplete penetration defect, which occurs due to insufficient penetration of the molten welding metal in the root pass. The factors that produce this defect are improper adjustment of welding parameters, improper connection design, and the use of back strips [10]. The location of this defect is shown schematically in **Figure 6**.

Generally, fracture of ductile steels characterized by the nucleation, growth and coalescence of microvoids [27, 43]. The welding defects has formed during the

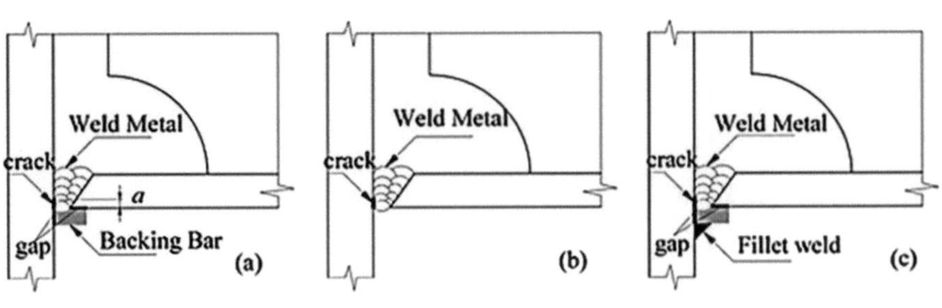

Figure 6.
Schematic of defects in beam-to-column welding connection. (a) Cracks in the beam-to-column connection with the presence of backing strip, (b) crack in connection without backing bar, (c) connection with backing bar and corner welding [10].

construction of steel structures act like a notch. The presence of the notch in a steel material creates the triaxial state of stress at the site of failure initiation. The triaxiality strongly affects the ductile fracture of steels. The triaxiality is measured as the ratio of hydrostatic tension or mean stress σ_m to von Misses effective stress σ_e [43]. Rice and Tracy showed that the rate of change of the mean hole radios R strongly depends upon the stress triaxiality [44]:

$$\frac{dR}{R} = 0.28d\varepsilon exp\left(\frac{3\sigma_m}{2\sigma_e}\right) \tag{8}$$

Where dR is the change in hole radius for a given increment of plastic strain dε. This means that the radius of voids on the fracture surface of a ductile steel increases with increasing triaxiality and the fracture surface will be similar to a brittle fracture mode [44].

In steel structures and in the conditions of dynamic loads of an earthquake, the simultaneous effect of notch and load application rate is encountered. Mirza et al. [43]. conducted a series of tensile test on notched milled steel samples at different strain rates and discussed the dependence of the failure mode on notch and strain rate. Fractographic examination of the fractured surfaces using scanning electron microscopy for 4 and 0.5 mm notch profiles tested at high rates (3×10^3 s^{-1}) and quasi static rates is shown in **Figure** 7. The 4 mm notch specimen for both strain rate shows a dimple fracture surfaces representing ductile fracture (**Figure 7a** and **c**) whereas for 0.5 mm notch (severe condition)the surface displays flat facet containing large multiple cracks, representing a brittle fracture mode (**Figure 4b** and **d**). Summary results of fractography examination for mild stees are given in **Table 1** [43]. Dependance of the fracture mechanism on the notch radios and strain rate can be clearly seen in the table. The results of the table show a stronger dependence of the failure mode on the notch severity compared to the strain rate.

By using the science of fracture mechanics in the design of structures, assuming the presence of cracks which is inevitable, the conditions can be advanced in such a way as to prevent the premature brittle failure and occurrence of severe losses caused by earthquakes [24, 45–47]. In the theory of fracture mechanics, a parameter is introduced as stress intensity factor, which has the following relationship with stress, σ, and defect length *a* [48, 49]:

$$K_I = f(g)\sigma\sqrt{\pi a} \tag{9}$$

f(g) is a function of the geometry of the part and the crack.

According to this criterion, failure occurs when the value of K_I reaches the critical value of K_{IC}. The K_{IC} is called fracture toughness of material. It is a material property in the same sense as that yield strength [50]. As mentioned above, the fracture toughness is also affected by the strain rate where its value decreases at high strain rates (**Figure 5**). The value of stress intensity factor can be measured using standard samples. In France, the NFA 03–180 standard specifies the necessary conditions for performing the required tests to determine the K_{IC} value of steel structures [35, 45, 51]. It is worth mentioning that steel structures are usually made of low carbon and low alloy steels. Using the critical stress intensity factor for this group of materials causes many errors. Therefore, this criterion can only be used for them at low temperatures and where the material is in a brittle state. To solve this problem, modeling and two-dimensional and

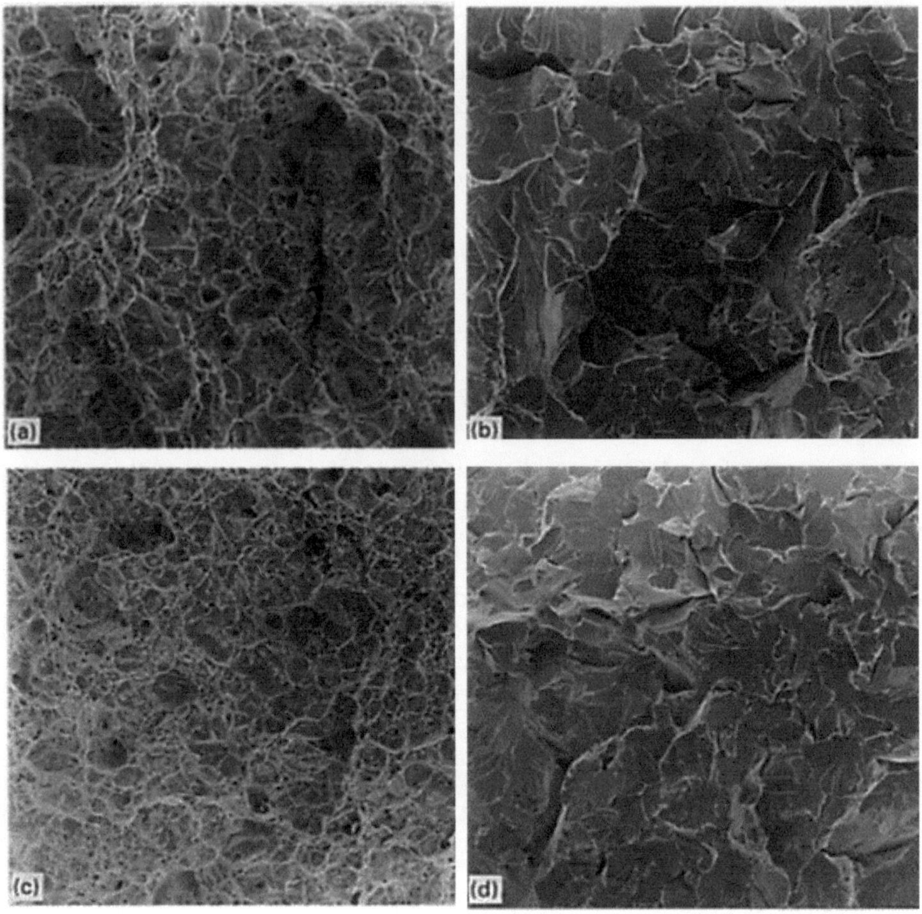

Figure 7.
SEM image of a mild steel, (a) and (b) 4 and 0.5 mm notch specimen tested at 10 ms^{-1} *(c) and (d) 4 mm and 0.5 mm notch specimen tested at quasi-static state [43].*

Notch radius(mm)	2.5×10^3 s^{-1}	3.5×10^2 s^{-1}	$8 \times 10^{-2} s^{-1}$ Quasi-static
4	D	D	D
2	B	D	D
1	B	B	D
0.5	B	B	B
D = ductile fracture; and B = brittle fracture.			

Table 1.
Variation of failure mode from ductile to brittle for a mild steel [43].

three-dimensional finite element methods have been used to evaluate the failure of structural steels and their connections in higher temperatures [24, 32, 47, 52].

In Eq. (9) determining the f(g) function is an important topic in fracture mechanics and has been determined for many parts with different geometries and cracks. In cases where this function is not known, the stress intensity factor can be determined

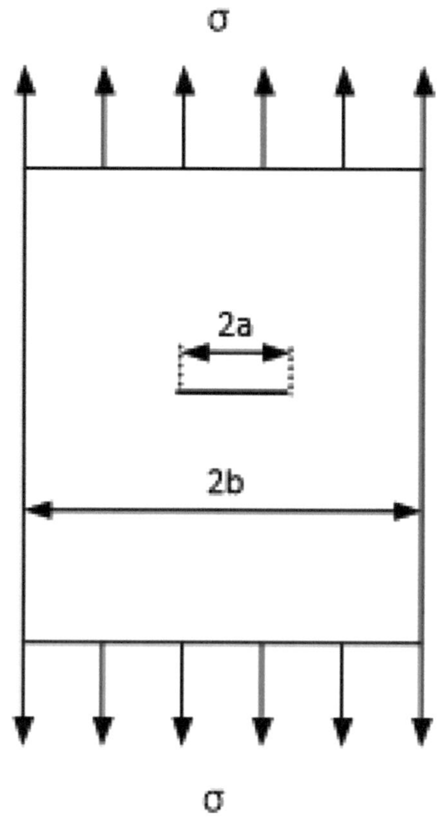

Figure 8.
A plate with a limited width has a central crack.

using numerical methods. Here are some situations that are investigated in the field of steel structures that are the subject of this research.

The stress intensity factor for a plate with a large width and a central crack of length 2a (shown in **Figure 8**) is expressed according to Eq. (10) [49].

$$K = \sigma\sqrt{\pi a} \tag{10}$$

If the width of the sheet is limited and equal to 2b, the correction factor should be considered so that the effect of the stress on the edges of the sample in the stress distribution of the crack tip can be included in the relationship of the stress intensity factor. In this case, to Eq. (10) will be as follows [49]:

$$K_I = \sigma\sqrt{\pi a}\left(\frac{2b}{\pi a}\tan\frac{\pi a}{2b}\right) \tag{11}$$

Connecting beam to column is a very common connection in steel structure buildings. Numerous defects in the joint have been the cause of cracking and failure of steel structures in recent earthquakes [6, 17–19, 53, 54]. Examples of these fractures are shown in **Figures 1** and **2**. Due to the complexity, the evaluation of the failure conditions of the beam-column joint should be done using finite element methods. Here one of the presented models is introduced. The model is based a type of

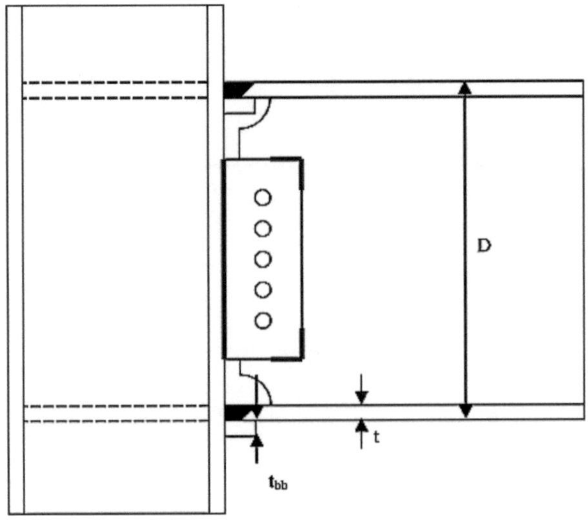

Figure 9.
A schematic of the beam-to-column connection design before the Northridge earthquake [22].

connection design used in structures before the Northridge earthquake [22]. This design method is shown in **Figure 9**. Investigations have shown many failures in the lower flange part of these connections. It was easily determined that the origin of these cracks was the lack of sufficient penetration of the weld, which occurred at the root pass with a small radius as shown in **Figure 10**. The radius of these fine defects is in the range of micrometers, which can be considered and modeled as a crack.

In this model, the crack of the bottom flange of the beam to the column is considered as an edge crack like the one shown in **Figure 11a**. The stress intensity factors are defined as the following relations depending on the geometry of the crack and the way of loading [22]:

$$K_t = \sigma_t \sqrt{\frac{\pi a}{Q}} Y_t \tag{12}$$

$$K_b = \sigma_b \sqrt{\frac{\pi a}{Q}} Y_b \tag{13}$$

where, parameters σ_t and σ_b are reference stresses for tension and bending, respectively. Y_t, Y_b and Q are stress magnification factors which are defined as the following relationships:

$$Y_t = 3.776 - 64.493\left(\frac{a}{t}\right) + 627.2\left(\frac{a}{t}\right)^2 - 2771\left(\frac{a}{t}\right)^3 + 6140\left(\frac{a}{t}\right)^4 - 6583\left(\frac{a}{t}\right)^5 + -2741\left(\frac{a}{t}\right)^6 \tag{14}$$

$$Y_b = 2.168 - 27.1\left(\frac{a}{t}\right) + 237.7\left(\frac{a}{t}\right)^2 1003\left(\frac{a}{t}\right)^3 + 2168\left(\frac{a}{t}\right)^4 - 2290\left(\frac{a}{t}\right)^5 + 944.8\left(\frac{a}{t}\right)^6 \tag{15}$$

$$Q = 1 + 1.464\left(\frac{a}{c}\right)^{1.65} \tag{16}$$

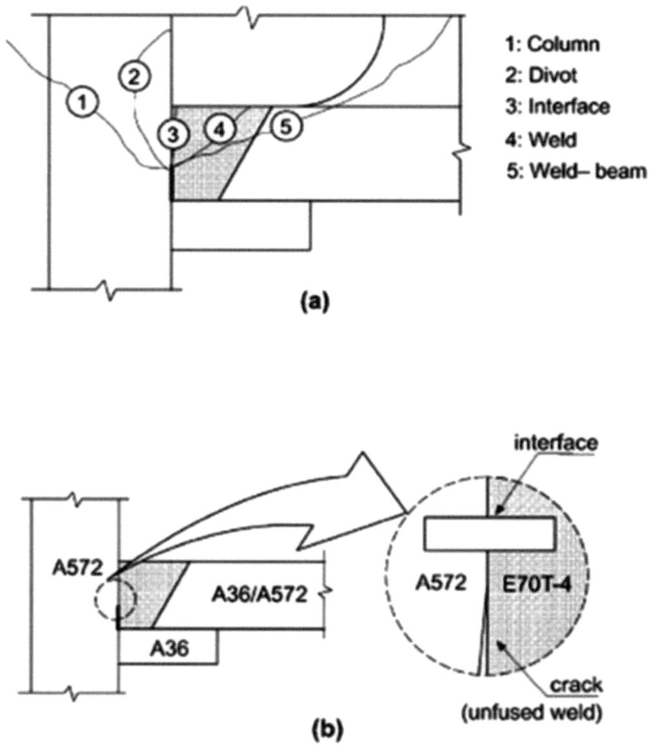

Figure 10.
(a) Schematic of beam and column failure of structures due to earthquake (The origin of the defect is lack of sufficient penetration), (b) Bimaterial interface at the weld–column fusion line [14].

where c is the half crack width. This particular dimension of the crack, which is not seen in **Figure 11a**, represents its out-of-plane dimension. It is worth mentioning that in the case that the external moment M is also applied, the total stress intensity factor is given as follow [22]:

$$K = \frac{M}{2I}\sqrt{\frac{\pi a}{Q}}[(D - t)Y_t + tY_b] \tag{17}$$

In this relation, I is the second moment of area of beam section, D is the depth and t is the flange thickness. In the meantime, when the back strip is used (**Figure 11b**), the conditions become more critical, so Yt and Yb are multiplied by the $\frac{Y_1}{1.12}$ factor. Where Y1 can be calculated from the following equation:

$$Y_1 = 1.081\left(\frac{\left(\frac{a}{t}\right)}{\left(\frac{a}{t}\right) + \left(\frac{t_b}{t}\right)}\right)^{-0.33} \tag{18}$$

4. Effects of welding residual stress

One of the main problems of metal structures is residual stress and distortion caused by localized heat during welding [26, 55]. Residual stresses are the stresses that

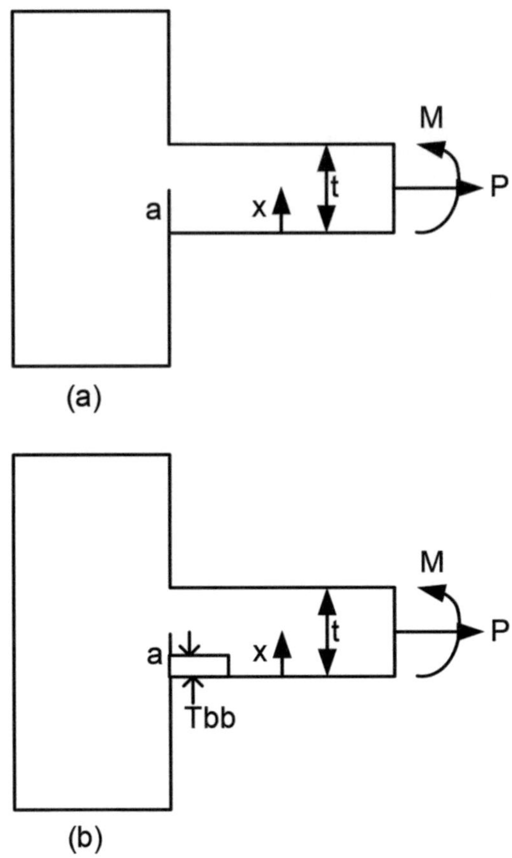

Figure 11.
(a) Schematic of beam-to-column connection under load P and moment M, and (b) showing the same conditions with the presence of a back strip.

exist in a body when external loads are removed. These stresses are also called internal stresses in some cases. The risk of brittle fracture in welded structures generally increases in the presence of residual stresses. Panontin et al. show that residual stress can increase the local stress triaxiality ahead of the axial crack front in a cylindrical pressure vessel [56]. The increased triaxiality caused by residual stress can transform a low-constraint structural configuration into a higher constraint (more sever) leading to cleavage fracture at a lower load level [56]. Zhang and Dong [57] and Matos and Dodds [23] found that residual stress alone consume approximately 30% of the available fracture toughness (about 65 MPa$\sqrt{m}$) for connection welds utilizing E70T-4 electrodes [21].

Residual stresses are caused by various factors such as mechanical work and welding in parts [22, 58, 59]. There are many factors that are effective in the formation of residual stresses. These parameters include design parameters, material parameters and manufacturing parameters. Design parameters include how to connect components. Material parameters include base metal and weld metallurgical conditions. Welding parameters include welding method, input heat, preheating, and how to restrain components during welding [60].

In welding processes two factors are determined. Firstly, the welding residual stress by cause of temperature changes during welding. Second, in some steel grades,

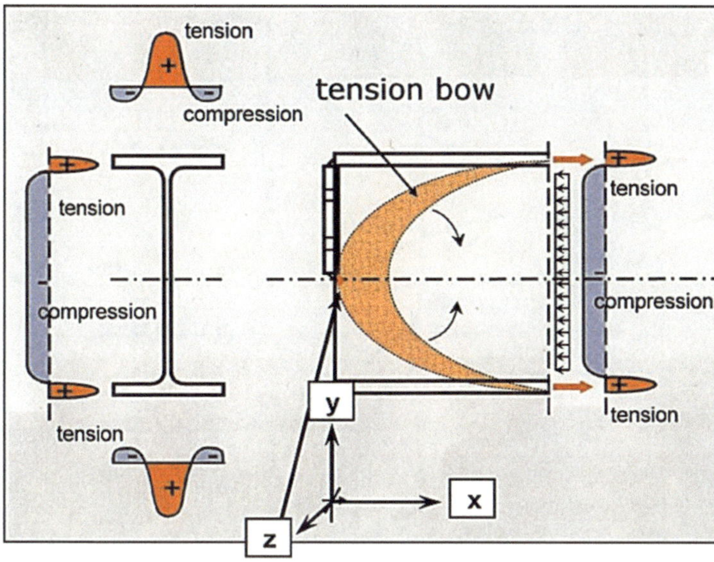

Figure 12.
Distribution of residual stresses in a rolled steel 'I' beam with a welded end-plate [63].

the solid-state austenite to martensite transformation in the time of cooling generates a significant value of the residual stresses [61, 62]. The former case is caused by non-uniform temperature changes in the body and are often known as thermal stresses. In fact, during the welding operation, the contraction and expansion of the weld metal and the base metal near the weld metal due to heating and cooling are restrained by the base metal away from the weld metal (**Figure 12**). Its result is the generation of tensile residual stresses in the weld and compressive residual stresses in the base metal away from the weld [63].

Residual stresses caused by welding are applied both longitudinally and transversely to the weld and the area affected by the weld. Meanwhile, the dimensions of the longitudinal residual stresses are wider. **Figure 13** shows a schematic representation of σ_x stress changes during welding according to temperature in different areas of a butt joint [63].

The most important negative effects of residual stresses in a welded joint are the addition of these stresses to the applied loads during service, which are usually not considered in the design [55, 64]. This multiplies the importance of studying the dimensions of these stresses in a joint [63].

The effect of residual stresses is modeled by Righiniotis et al. [24]. The model is based on the fact that the residual stresses are applied uniformly and to the dimensions of half the yield point in the thickness of the flange, where σ_y is the material's yield strength. According to this, the residual stress filed is express as follow [20]:

$$\sigma_{res} = \sigma_y(1 - 0.917u - 14.533u^2 + 83.115u^3 - 215.45u^4 + 244.16u^5 - 96.36u^6 \quad (19)$$

Where $u = x/t$, x and t parameter are present in **Figure 11a**.

The stress intensity factor (K_{res}) for this type of crack face loadings is developed utilizing the weight function method [65]. Based on this, the stress intensity factor for an arbitrary crack is expressed as follows:

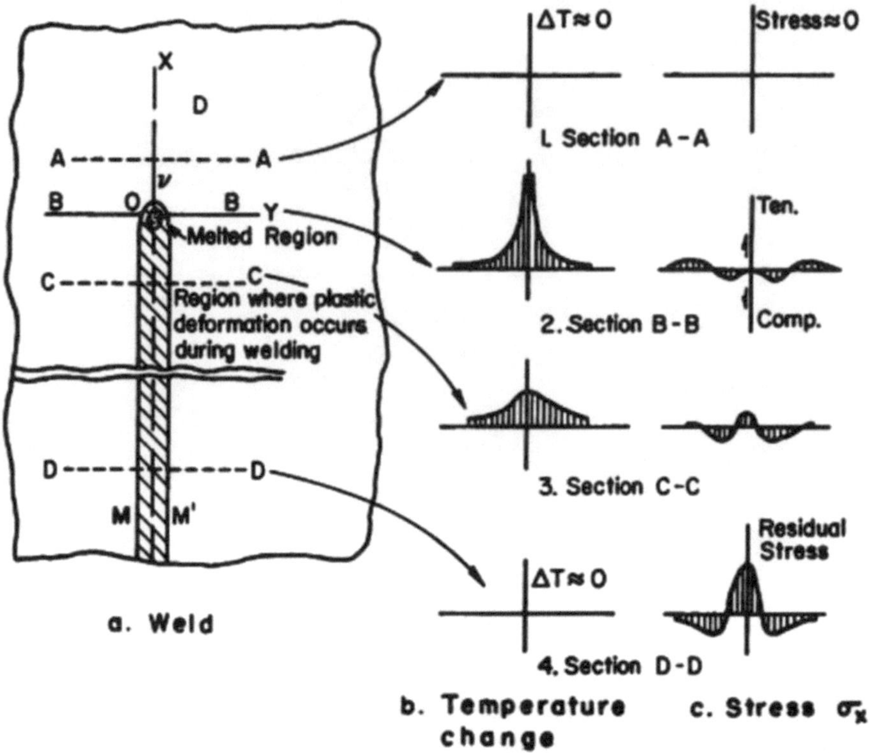

Figure 13.
Longitudinal residual stress changes according to temperature in a butt joint [63].

$$K = \int_0^a \sigma(x)h(a,x)dx \tag{20}$$

Where $h(a,x)$ is the weight function and given as:

$$h(a,x) = \frac{E}{(1-u^2)K^*(a)} \frac{\partial u^*(a,x)}{\partial a} \tag{21}$$

In Eq. (20), for plan strain state, u is the crack opening displacement, E is the Young's modulus and u denotes to Poisson's ratio. The * parameters K (a) refers to a known reference loading condition.

Although the reference stress intensity factor is known for many cracks with different geometries [66], not much information is available about the reference displacement solution. For such cases, the weight function is expressed as an infinite series with undetermined coefficients. For example, for an edge cracks the weight function may be written as [65]:

$$h(a,x) = \sqrt{\frac{2}{\pi a}}\left[\frac{r}{\sqrt{1-r}} + \sum_{n=0}^{\infty} C_n(1-r)^{n+\frac{1}{2}}\right] \tag{22}$$

where $r = x/a$ as shown in **Figure 11a**. The self-consistency and the geometry conditions are method to evaluate some of coefficients in Eq. (22) [65].

The self-consistency cannot be used in this state due to non-symmetric geometry of **Figure 11a**. The geometry condition state that using of Eq. (20) to different loading conditions should result in their respective stress intensity factors.

In the case presented here, two constants C_n can be calculated using the solution for tension, Y_t, and bending, Y_b. By determination of C_0 and C_1 and substitution in Eq. (22) and using Eqs. (20) and (21) results in the stress magnification factor for any other loading $\sigma(u)$ as:

$$Y(v) = \frac{\sqrt{2}}{\pi v}\sigma(u)\left\{\frac{u\left[1-\left(\frac{u}{v}\right)\right]^{-\frac{1}{2}}}{v} + C_0\left(1-\frac{u}{v}\right)^{\frac{1}{2}} + C_1\left(1-\frac{u}{v}\right)^{\frac{3}{2}}\right\}\cdot du \tag{23}$$

For many cracks, when faced with loading conditions such as residual stresses that are distributed in the thickness, it is expressed a nominal function (for example Eq. (11)) [22]. Here the magnification factor will be given as

$$Y = \sum_{i=0}^{n} A_i Y_i \tag{24}$$

Where Y_i is related to u_i. Determination of the stress magnification factor in this case, Y_{res}, is associated with determination of Y_i and weighting their contribution as prescribed by Eq. (24) [22]. By determining the stress magnification factor, the value of the stress intensity factor related to the residual stress is stated as follows

$$\mathrm{K}_{res} = \sigma_y\sqrt{\frac{\pi \mathrm{a}}{\mathrm{Q}}}\mathrm{Y}_{res} \tag{25}$$

For a fully elastic flange, fracture moment is expressed as:

$$\mathrm{M}_f = 2(K_{IC} - \mathrm{K}_{res})\sqrt{\frac{\mathrm{Q}}{\pi \mathrm{a}}}I[D-t)\mathrm{Y}_t + t\mathrm{Y}_b]^{-1} \qquad M \leq \mathrm{M}_y \tag{26}$$

Where K_{IC} is the material's plane strain fracture toughness and M_y is the beam's yield moment. The inequality of 26 indicates the upper bond of the Linear Elastic Fracture Mechanics (LEFM) credit.

A common solution to reduce residual stress is to perform a post weld heat treatment, PWHT, after part fabrication which helps to homogenize the microstructure. It was found that PWHT can reduce residual stress up to 70% when the treatment is performed in the 600–700°C range [67].

5. Conclusions

The extensive failures of steel structures during the Northridge and Kobe earthquakes showed that the available information regarding the behavior of materials used in the construction of steel structures and also their design has many deficiencies in earthquake loading conditions. This is not far-fetched considering the basis of the information obtained from the classical plasticity theory, which is based on the static tensile test. Comprehensive field and laboratory investigations were conducted by manufacturers, consumers, designers and welding inspectors on the remains of destroyed structures during these two earthquakes. The investigations have shown that three important factors in the brittle failure of structures are high strain rate, welding defects, welding residual stress. In general, the yield stress of structural steels increases with increasing the strain rate where the lower strength materials exhibits the most strain sensitivity compares with higher strength steels grades. The increasing strain rate also decreases the fracture toughness of materials. This means that the transition of ductile into brittle takes place at higher temperatures. Welding defects are among the important factors of structural failure under earthquake loads. These impefections created during the construction of steel structures act like cracks. Cracks facilitate brittle fracture of steels. In this chapter, this issue is well analyzed from the viewpoint of fracture mechanics. The welding process also induces a local triaxial residual stress field in the welding connections. The order of residual stress can be in order of the yield point of the steels and should be considered in design and manufacturing of steel structuers (for example, performing stress relief heat treatments).

Author details

Alireza Khalifeh[1*], Habib Danesh Manesh[2] and Abolghasem Dehghan[2]

1 Shiraz Petrochemical Complex, Shiraz, Iran

2 Department of Materials Science and Engineering, School of Engineering, Shiraz University, Shiraz, Iran

*Address all correspondence to: ar.khalifeh@spc.co.ir

References

[1] Porter K, Shoaf K, Seligson H. Value of injuries in the Northridge earthquake. Earthquake Spectra. 2006; **22**(2):555-563

[2] Aoki N et al. Survival and cost analysis of fatalities of the Kobe earthquake in Japan. Prehospital Emergency Care. 2004;**8**(2):217-222

[3] Clifton C et al. Steel structures damage from the Christchurch earthquake series of 2010 and 2011. Bulletin of the New Zealand Society for Earthquake Engineering. 2011;**44**(4): 297-318

[4] Bruneau M et al. Preliminary report on steel building damage from the Darfield earthquake of September 4, 2010. Bulletin of the New Zealand Society for Earthquake Engineering. 2010;**43**(4):351-359

[5] Horikawa K, Sakino Y. Damages to steel structures caused by the 1995 Kobe earthquake. Structural Engineering International. 1996;**6**(3): 181-182

[6] Tremblay R, Filiatrault A, Timler P, Bruneau M. Performance of steel structures during the 1994 Northridge earthquake. Canadian Journal of Civil Engineering. 1995;**22**(2):338-360

[7] MacRae G et al. Lessons from steel structures in Christchurch earthquakes. In: Proceedings of the 8th International Conference on Behavior of Steel Structures in Seismic Areas (STESSA). Shanghai, China: Tongji University; 2015. pp. 1474-1481

[8] Muguruma H, Nishiyama M, Watanabe F. Lessons learned from the Kobe earthquake—A Japanese perspective. PCI Journal. 1995;**40**(4):28-42

[9] Miller DK. Lessons learned from the Northridge earthquake. Engineering Structures. 1998;**20**(4–6):249-260

[10] Wang Y, Zhou H, Shi Y, Chen H. Fracture behavior analyses of welded beam-to-column connections based on elastic and inelastic fracture mechanics. International Journal of Steel Structures. 2010;**10**:253-265

[11] Rolfe ST, Barsom JM. Fracture and Fatigue Control in Structures: Applications of Fracture Mechanics. USA: ASTM International; 1977

[12] Venture SJ. Interim guidelines: Evaluation, repair, modification and design of steel moment frames. Report No. SAC. 1995;**95**:2

[13] Kaufmann EJ, Fisher JW. Fracture analysis of failed moment frame weld joints produced in full scale laboratory tests and buildings damaged in the Northridge earthquake. In: Experimental Investigations of Materials, Weldments and Nondestructive Examination Techniques. Technical Report No SAC 95-08. Sacramento, California: SAC Joint Venture; 1995

[14] Matos C, Dodds R Jr. Probabilistic modeling of weld fracture in steel frame connections part I: Quasi-static loading. Engineering Structures. 2001;**23**(8): 1011-1030

[15] Kaufmann E, Fisher J, Di-Julio-Jr R, Gross J. Failure analysis of welded steel moment resisting frames damaged in the northridge earthquake. In: Technical Report 5625. USA: NIST; 1995

[16] Youssef NF, Bonowitz D, Gross JL. A Survey of Steel Moment-Resisting Frame Buildings Affected by the 1994 Northridge Earthquake. Gaithersburg,

MD, USA: US National Institute of Standards and Technology; 1995

[17] Mahin SA. Lessons from damage to steel buildings during the Northridge earthquake. Engineering Structures. 1998;**20**(4–6):261-270

[18] Popov EP, Yang T-S, Chang S-P. Design of steel MRF connections before and after 1994 Northridge earthquake. Engineering Structures. 1998;**20**(12): 1030-1038

[19] Kuwamura H. Fracture of steel during an earthquake—State-of-the-art in Japan. Engineering Structures. 1998; **20**(4–6):310-322

[20] Standard B. Guide on Methods for Assessing the Acceptability of Flaws in Fusion Welded Structures. 1991;**BS7910**: 6393-6493

[21] Kaufman E, Fisher J. A study of the effects of material and welding factors on moment frame weld joint performance using a small scale tension specimen. SAC Joint Venture. 1995:95-108

[22] Righiniotis T, Imam B. Fracture reliability of a typical Northridge steel moment resisting connection. Engineering Structures. 2004;**26**(3): 381-390

[23] Matos C, Dodds R Jr. Modeling the effects of residual stresses on defects in welds of steel frame connections. Engineering Structures. 2000;**22**(9): 1103-1120

[24] Righiniotis T, Omer E, Elghazouli A. A simplified crack model for weld fracture in steel moment connections. Engineering Structures. 2002;**24**(9): 1133-1140

[25] Righiniotis TD, Hobbs RE. Fracture strength of a moment resisting welded connection under combined loading: Part II—Results. Journal of Constructional Steel Research. 2000; **56**(1):31-45

[26] Khalifeh A. Stress corrosion cracking behavior of materials. Engineering Failure Analysis. 2020;**10**:55-75

[27] Khalifeh A, Banaraki AD, Manesh HD, Banaraki MD. Investigating of the tensile mechanical properties of structural steels at high strain rates. Materials Science and Engineering: A. 2018;**712**:232-239

[28] Miura K et al. High Strain Rate Deformation of High Strength Sheet Steels for Automotive Parts. SAE Technical Paper; 1998

[29] Hosford WF. Mechanical Behavior of Materials. USA: Cambridge University Press; 2010

[30] Chang K-C, Lee GC. Strain rate effect on structural steel under cyclic loading. Journal of Engineering Mechanics. 1987;**113**(9):1292-1301

[31] Kaufmann E, Fisher J, DiJulio R, Gross J. Failure Analysis of Welded Steel Moment Frames Damaged in the Northridge Earthquake (NISTIR 5944). Gaithersburg, MD: National Institute of Standards and Technology; 1997

[32] Matos C, Dodds R Jr. Probabilistic modeling of weld fracture in steel frame connections part II: Seismic loading. Engineering Structures. 2002;**24**(6): 687-705

[33] Manjoine M. Influence of rate of strain and temperature on yield stresses of mild steel. Journal of Applied Mechanics. 1944;**11**(4):A211-A218

[34] Hull D, Bacon DJ. Introduction to Dislocations. Vol. 37. UK: Elsevier; 2011

[35] Blondeau R. Metallurgy and Mechanics of Welding: Processes and Industrial Applications. UK: John Wiley & Sons; 2013

[36] Srinivas M, Kamat S. Effect of strain rate on fracture toughness of mild steel. Materials Science and Technology. 2001; **17**(5):529-535

[37] Zhu H, Zhang C, Chen S, Wu J. A modified Johnson-cook constitutive model for structural steel after cooling from high temperature. Construction and Building Materials. 2022;**340**:127746

[38] Lu J, Liu H, Chen Z, Liao X. Experimental investigation into the post-fire mechanical properties of hot-rolled and cold-formed steels. Journal of Constructional Steel Research. 2016;**121**: 291-310

[39] He A, Xie G, Zhang H, Wang X. A comparative study on Johnson–cook, modified Johnson–cook and Arrhenius-type constitutive models to predict the high temperature flow stress in 20CrMo alloy steel. Materials & Design (1980–2015). 2013;**52**:677-685

[40] Niu Q-l et al. Dynamic mechanical behavior of ultra-high strength steel 30CrMnSiNi2A at high strain rates and elevated temperatures. Journal of Iron and Steel Research International. 2017; **24**(7):724-729

[41] Gates W, Morden M. Lessons from inspection, evaluation, repair and construction of welded steel moment frames following the Northridge earthquake. Surveys and Assessment of Damage to Buildings Affected by the Northridge Earthquake of January. 1995; **17**:1994

[42] Bonowitz D, Youssef N. SAC survey of steel moment frames affected by the 1994 Northridge earthquake. Surveys and Assessment of Damage to Buildings Affected by the Northridge Earthquake of January. 1994;**17**:95-06

[43] Mirza M, Barton D, Church P. The effect of stress triaxiality and strain-rate on the fracture characteristics of ductile metals. Journal of Materials Science. 1996;**31**:453-461

[44] Rice JR, Tracey DM. On the ductile enlargement of voids in triaxial stress fields∗. Journal of the Mechanics and Physics of Solids. 1969;**17**(3):201-217

[45] Lewinsohn CA. Mechanical Behavior of Materials by Norman E. Dowling. USA: Taylor & Francis; 2000

[46] Chi W-M, Deierlein GG, Ingraffea A. Fracture toughness demands in welded beam-column moment connections. Journal of Structural Engineering. 2000; **126**(1):88-97

[47] Chi W-M, Deierlein GG, Ingraffea AR. Finite-element fracture analyses of welded beam-column connections. ASTM Special Technical Publication. 2000;**1360**:439-455

[48] Kirkemo F. Applications of probabilistic fracture mechanics to offshore structures. Applied Mechanics Reviews. 1988;**41**(2):61-84

[49] Shih G. Handbook of Stress-Intensity Factors for Researchers and Engineers. Bethlehem, PA. THE THE AAL D1 D2 D3 ABS FNC K1LE K1LE: Institute of Fracture and Solid Mechanics, Lehigh University; 1973

[50] Dieter GE, Bacon D. Mechanical Metallurgy. Vol. 3. New York: McGraw-Hill; 1976

[51] Bousseau M. Fracture Toughness of Welded Joints. Metallurgy and

Mechanics of Welding: Processes and Industrial Applications, USA: John Wiley & Sons; 2008. pp. 239-277

[52] Wang Y, Zhou H, Shi Y, Xiong J. Fracture prediction of welded steel connections using traditional fracture mechanics and calibrated micromechanics based models. International Journal of Steel Structures. 2011;**11**:351-366

[53] Venture SJ, Committee GD. Recommended Seismic Evaluation and Upgrade Criteria for Existing Welded Steel Moment-Frame Buildings. Washington, DC: Federal Emergency Management Agency; 2000

[54] Nakashima M, Inoue K, Tada M. Classification of damage to steel buildings observed in the 1995 Hyogoken-Nanbu earthquake. Engineering Structures. 1998;**20**(4–6): 271-281

[55] Khalifeh A, Banaraki AD, Daneshmanesh H, Paydar M. Stress corrosion cracking of a circulation water heater tubesheet. Engineering Failure Analysis. 2017;**78**:55-66

[56] Panontin T, Hill M. The effect of residual stresses on brittle and ductile fracture initiation predicted by micromechanical models. International Journal of Fracture. 1996;**82**:317-333

[57] Zhang J, Dong P. Residual stresses in welded moment frames and implications for structural performance. Journal of Structural Engineering. 2000;**126**(3): 306-315

[58] Masubuchi K. Analysis of Welded Structures: Residual Stresses, Distortion and their Consequences. Oxford: Pergamon Press; 1980

[59] Huang T, Dong P, DeCan LA, Harwig DD. Residual stresses and distortions in lightweight ship panel structures. Technology Review Journal. 2003;**11**(1):1-26

[60] Deng D. FEM prediction of welding residual stress and distortion in carbon steel considering phase transformation effects. Materials & Design. 2009;**30**(2): 359-366

[61] Lienert T, Babu S, Siewert T. Solid-state transformations in weldments. In: Acoff VL, editor. The ASM handbook®. Welding Fundamentals and Processes. Vol. 6. USA: Amazon; 2011. pp. 133-135

[62] Leggatt R. Residual stresses in welded structures. International Journal of Pressure Vessels and Piping. 2008; **85**(3):144-151

[63] Kou S. Welding metallurgy. New Jersey, USA. 2003;**431**(446):223-225

[64] Khalifeh A. Stress corrosion cracking damages. In: Failure Analysis. London, UK: IntechOpen; 2019

[65] Fett T, Munz D. Stress Intensity Factors andWeight Functions. UK: Computational Mechanics Publications; 1997

[66] Murakanal Y. Stress Intensity Factors Handbook. Vol. 1-2. Oxford, UK: Pergamon Press; 1987

[67] Aba-Perea P, Pirling T, Preuss M. In-situ residual stress analysis during annealing treatments using neutron diffraction in combination with a novel furnace design. Materials & Design. 2016;**110**:925-931

Chapter 4

Seismic Assessment of RC Shear Wall Structure under Real Seismic Sequences Using Equivalent Shell-Wire Model

Omid Karimzade Soureshjani and Ali Massumi

Abstract

Shear walls are commonly used in areas with moderate to high seismic hazards. They are cost-effective, simple, and significantly reduce the relative displacement. Accurate non-linear assessments of shear wall structures require highly detailed models and finite element (FE) analyses. The inclusion of analyses on the cumulative damage and damage evolution like seismic sequences adds a higher level of complexity to the assessments. Here, an equivalent shell-wire model is used to propose a simple, accurate technique with less computational cost. The areas under the capacity curves were modeled using 3D solid and equivalent shell-wall models and revealed only a 4.48% difference with a 50% difference in computational cost. The hysteresis curves of the experimental and equivalent shell-wire models showed good agreement. To examine the cumulative damage, damage evolution, strength and stiffness degradation of the model, a six-story structure was designed and analyzed under seven mainshocks-aftershocks in both cases of 3D solid and equivalent shell-wire models. The results showed that increasing the plastic strain or damage increased the difference between the results of the models, but this was negligible. Besides the accurate and appropriate results of the equivalent shell-wire model, it required about 39% less computational cost than the 3D solid model.

Keywords: mainshock-aftershock, shear wall, equivalent shell-wire model, cumulative damage, nonlinear seismic behavior

1. Introduction

Concrete shear walls are reinforced vertical structural elements that are designed to resist lateral forces, increase structural stiffness, and prevent high inter-story drift ratios [1]. Because of the relatively simple construction procedure, high seismic performance, high energy dissipation, and ductile behavior, shear wall systems have become popular for use in mid- to high-rise structures in cities with moderate-to-high seismic hazards [2, 3]. This makes accurate seismic behavior by shear-wall structures of importance.

Shear walls in structural systems should be able to tolerate either axial or lateral loads to increase the structural system performance [4]. Study of the seismic behavior of such structures in past earthquakes has shown that they are susceptible to flaws in the design procedure. In cases such as in Nicaragua (1972), Mexico (1985), and Armenia (1988), the lack of an appropriate design caused shear-wall collapse [2, 5]. Shear walls should exhibit ductile behavior under large structural forces without failing [6]. Design problems, the location of the shear walls, and openings in the plans as well as construction problems can decrease the efficiency of the shear walls under seismic loads and should be accurately considered during the design [7].

The behavior of shear wall structures under seismic loading has been investigated in a variety of studies. However, these generally have focused on single-degree-of-freedom (SDOF) systems or 2D frames of low height. Generally, 3D structures have been neglected because of the complexity of analysis, design factors, and the computational cost (simply, the computational cost is the time required to perform an analysis) [8, 9, 10].

Experimental studies also have usually been limited to samples having low heights using simple cyclic tests [11]. This could be because of the complex behavior, size of the shear-wall elements, and the interactions between the concrete, steel rebars, shear wall, and main structure in the 3D solid models that increase the modeling complexity and computational cost. In some previous studies, structures have exhibited different seismic behaviors under seismic sequences depending on the use of 2D or 3D modeling; thus, the 3D models have been preferred [12]. The lack of an appropriate technique for modeling shear wall structures at less computational cost with accurate results is evident. This study used an equivalent shell-wire technique to decrease the computational cost in the 3D modeling of shear-wall structures having high accuracy.

A mainshock is usually accompanied by a group of ground motions. Such aftershocks can increase structural damage and cause failure and even building collapse [13, 14]. The causes for these include damage evolution, damage accumulation, strength, and stiffness degradation under seismic sequences or mainshock-aftershocks. Many studies have evaluated the effect of the seismic sequences on the structures; however, because of the high computational cost and analysis complexity, they usually have been limited to short structures or 2D models.

Studying the seismic behavior of the shear wall structures under seismic sequences could be a big challenge that required high computational costs. The complexity of the structural elements and the shear wall, the size of the shear wall elements and their interactions as well as consideration of the cumulative damage, damage evolution, strength, and stiffness degradation under seismic sequences increases the complexity of studying the seismic behavior of shear wall structures. The shell-wire model is a possible alternative to reducing the computational cost of such analyses. To evaluate the accuracy of the shell-wire technique over 3D solid models, it was compared with a full-scale experimental model under cyclic loading. This model also was investigated by design analysis of a six-story medium-height shear wall structure under real seismic sequences.

An alternative analytical model with appropriate accuracy (equivalent shell-wire model) is presented to analyze shear-wall structures. This model can consider damage evolution, cumulative damage and strength and stiffness degradation under arbitrary dynamic loading at a less computational cost compared to usual analytical models such as 3D solid models. The approach makes it possible to analyze taller structural models of greater complexity or with finer mesh to increase accuracy compared to a 3D solid model. **Figure 1** lists the aspects of this chapter.

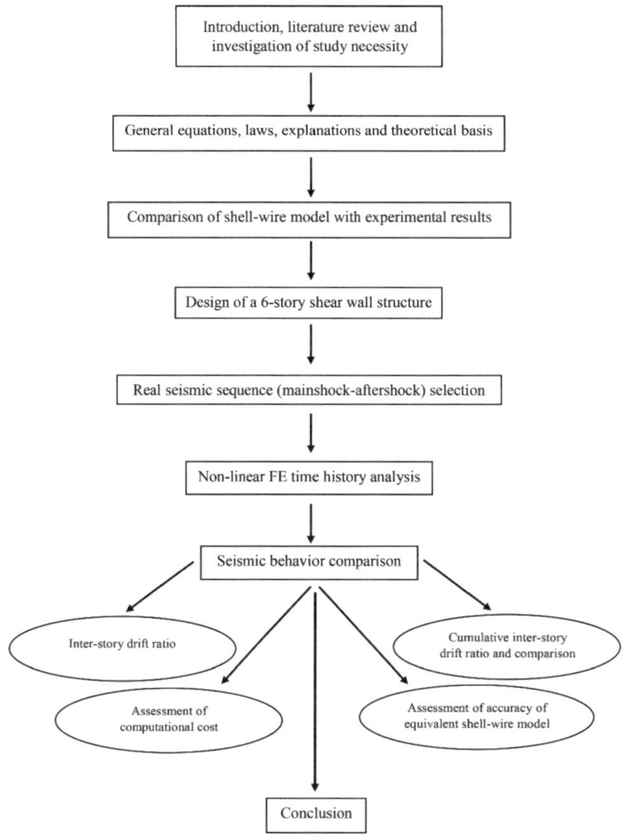

Figure 1.
Research strategy.

2. General equations, laws, explanations, and theoretical basis

2.1 3D shell element

It is possible to model structural elements using 3D solid or shell elements; however, the selection between 2D or 3D shell elements depends on the relative thickness of the structure. In other words, the third dimension could be neglected. Although 3D elements can be suitable for any geometry, for shell elements, the ratio of the thickness to the other dimensions should not be greater than 1 to 10 or less than 1 to 1000 [15]. For shear walls, the ratio of the thickness to height and width is the same for shell elements; thus, it is possible to consider the shear walls as shells or 3D solid elements.

The selected shell element was a 3D shell with a mesh element type of 4-node doubly curved general-purpose shell, reduced integration with hourglass control, and finite membrane strains (S4R) [16, 17]. This is a robust quadrilateral element for general-purpose applications and is suitable for thick and thin walls depending on the thickness of the elements. Each node has three degrees of freedom, as for 3D solid elements (**Figure 2**). It should be noted that because of the equality of the degree of freedom, it is easy to combine and tie a 3D shell (shear wall) and 3D solid elements (beams and columns). Thus, the 3D shell was selected.

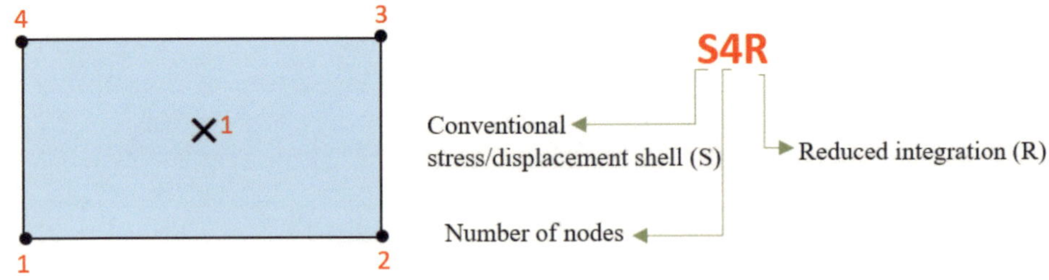

Figure 2.
The S4R element and integration points.

2.2 Wire element

Concrete shear walls contain concrete and rebars. In order to reduce the computational cost, the steel rebars were modeled as 3D wire elements. Ideally, wire elements are used when both their thicknesses and depths are considered to be small compared to their lengths. A 2-node linear 3D truss mesh element (T3D2) was considered for the wires that only support axial loading along the element [18]. It has been assumed that the steel rebars do not support the moments or forces perpendicular to the centerline.

It should be noted that because of the degree of freedom equality (shell shear wall, 3D solid columns, and beams), the 3D wire element was selected.

2.3 Shell-wire model

As the goal was the analysis of shear walls at less computational cost, the alternative model should be able to consider the cumulative damage, damage evolution, strength, and stiffness degradation under arbitrary loads. These results must be in good agreement with the complex 3D solid and experimental models. The equivalent shell-wire model considers shear walls as a combination of a shell (concrete) and wires (rebars) and uses equivalent concrete and rebars to provide accurate results and decrease the computational cost. **Figure 3a** and **b** show a simple 3D solid model and an equivalent shell-wire model, respectively.

2.4 Finite element analysis

The type of time integration method and material modeling have an intensive effect on the accuracy of the analysis [19]. For an accurate dynamic analysis, the time integration method should be chosen in accordance with the numerical dissipation, dispersion, wave propagation, excitation frequency content, convergence, desirable behavior and accuracy of the model, and computational cost [16]. Considering the frequency content of the excitation (earthquake acceleration in this case), desirable accuracy, and structural models the implicit dynamic time integration with the application of moderate dissipation was chosen for this study. This analysis covers more general dynamic events with appropriate accuracy [16, 18].

ABAQUS software was used to perform dynamic implicit analysis. Nonlinear implicit time-history analysis has been employed based on the unconditionally stable Hilber Hughes Taylor (HHT) method, with appropriate time increments and a full

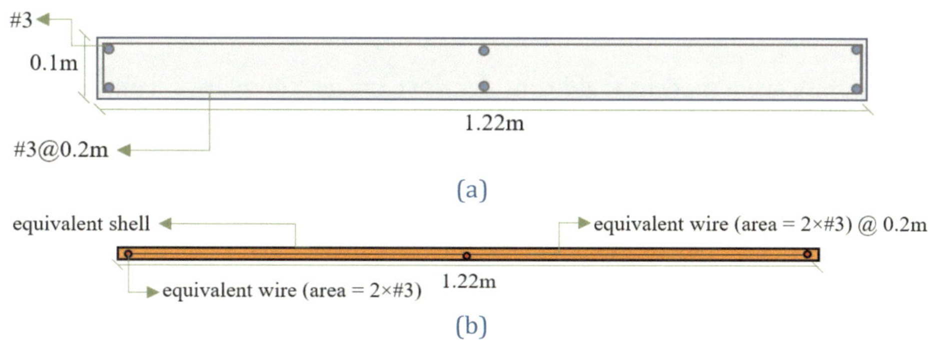

Figure 3.
Shear wall section: (a) 3D solid model; (b) equivalent shell-wire model scheme.

Newton–Raphson solution [18, 20, 21]. The size of time increments is fully automatic and is calculated based on an algorithm, rate of plasticity, and convergence [18]. The HHT method is a second-order accurate and unconditionally stable method that showed appropriate accuracy and convergence in general dynamic problems [22]. The numerical implementation of the time integration method is explained in Eqs. (1)–(8) [18, 23, 24].

The concrete damage plasticity model was used to define and model the concrete behavior. This model is based on the assumption of isotropic damage and is an analytical model used for the analysis of materials with concrete-like properties subjected to arbitrary loads [18]. Also, this model has a good ability to consider the interaction between rebars and concrete like the current study.

It should be noted that the ability to accurately consider the damage and damage evolution is one of the characteristics of the material model (Section 2.5).

The answer for every chosen time step and at any desired degree of freedom can be obtained as:

$$\Delta U|\tau = \tau^3 \Delta U|_{t+\Delta t} + \tau(1-\tau^2)\Delta t \dot{U}|_t + \tau^2(1-\tau)\frac{\Delta t^2}{2}\ddot{U}|t \tag{1}$$

$$\dot{U}|_t = \frac{\Upsilon}{\beta\tau\Delta t}\Delta U|\tau + \left(1-\frac{\Upsilon}{\beta}\right)\dot{U}|_t + \left(1-\frac{\Upsilon}{2\beta}\right)\tau\,\Delta t\,\ddot{U}|t \tag{2}$$

$$\ddot{U}|\tau = \frac{1}{\beta\tau^2\Delta t^2}\Delta U|_\tau - \frac{1}{\beta\tau\Delta t}\dot{U}|_t + \left(1-\frac{1}{2\beta}\right)\ddot{U}|t \tag{3}$$

$$\Delta U|\mathrm{t}+\Delta\mathrm{t} = U\,|\mathrm{t}+\Delta\mathrm{t} - U| \tag{4}$$

$$\beta = \frac{1}{4}(1-\alpha)^2 \tag{5}$$

$$\Upsilon = \frac{1}{2} - \alpha \tag{6}$$

$$-\frac{1}{2} \leq \alpha \leq 0 \tag{7}$$

$$0 \leqslant \tau \leqslant 1 \tag{8}$$

where $\ddot{U}$ is acceleration in an implicit dynamic step, $\dot{U}$ is the velocity, and U is the displacement for a degree of freedom.

2.5 Cumulative damage (damage evolution)

A comprehensive model should consider the damage, cumulative damage, damage evolution, strength, and stiffness degradation for the concrete and steel rebars. Herein, the damage and damage evolution was characterized by the degradation in strength and stiffness of the materials. The initial damage criteria of the material (concrete or steel rebars) depended on the specific strength. For ductile material (steel rebars), when the stress reached its limits, the evolution of the damage variable took an exponential form according to the fraction strain, strain rate, and displacement at failure [18, 25–27]. Experimental studies have shown that brittle material (concrete) under either compression or tension, the initial damage, damage evolution, or cumulative damage could be linked to the model provided by Lubliner et al. [28]. Thus, it was assumed that the compressive stiffness recovered entirely when the load changed from tension to compression. Under tensile loading, however, the stiffness did not recover because of the existence of small cracks [25, 26, 29, 30].

3. Experimental test

To investigate the accuracy of the shell-wire equivalent model, the capacity curve and cyclic behavior of a shear wall modeled using an equivalent shell-wire model were compared to the results of complex 3D solid and experimental models. The computational costs of the models also were compared. The accuracy of the 3D solid model has been previously investigated and the results have shown good agreement between the complex 3D solid and experimental models [25].

3.1 Capacity curve

The experimental shear wall used was the RW1 model as tested by Thomson and Wallace [11], the details of which are shown in **Figure 4a**. According to previous studies, there is a good agreement between the experimental test and the 3D solid model [25]. The capacity curves of the 3D solid model and the equivalent shell-wire were obtained using monotonic load and nonlinear FE analysis (Δtarget = 6.5 cm). **Figure 4b** shows the equivalent shell-wire model of the RW1 shear wall.

Figure 5 compares the accuracy of the equivalent shell-wire model by comparing the area under the displacement-base shear force curves (capacity curves), which are

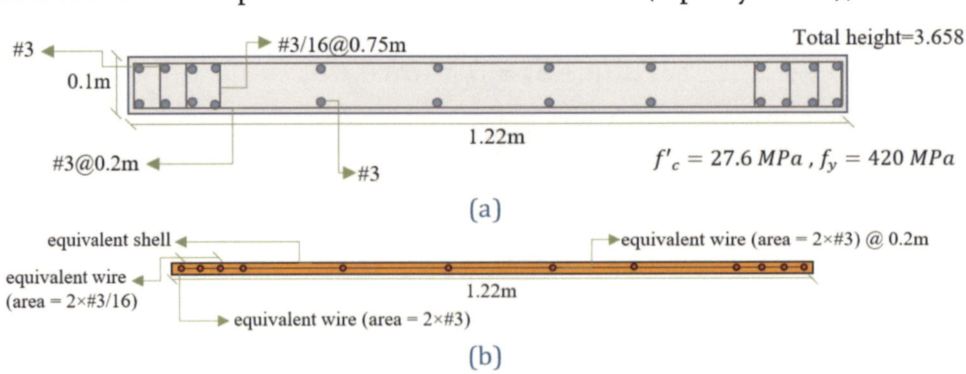

Figure 4.
RW1 model: (a) 3D solid and experimental models [25]; (b) equivalent shell-wire model scheme.

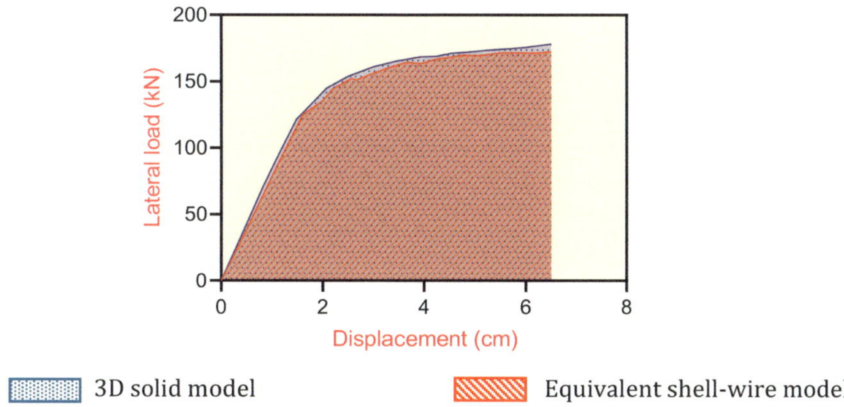

Figure 5.
Capacity curve comparison. 3D solid model. Equivalent shell-wire model.

equal to the absorbed energy. The models (equivalent shell-wire and 3D solid models) differed by only 4.48%, which indicates good agreement between them. The equivalent shell-wire model required about 50% less time to complete this task, which suggests that the computational cost also decreased by 50% when using the equivalent shell-wire model compared to the 3D solid model. It should be noted that the size of the mesh elements was the same for both models.

3.2 Cyclic behavior

The cyclic behavior of the equivalent shell-wire model was assessed to obtain a hysteresis loop using a displacement protocol. Hysteresis curves provide valuable information about the nonlinear behavior of the models under cyclic loading, including the strength and stiffness degradation [31–33]. Comparison of the hysteresis curves of the equivalent shell-wire model with the 3D solid and experimental models showed good agreement on both the compressive and tensile sides (**Figure 6**). This indicates that the equivalent shell-wire is an appropriate model that considers the strength and stiffness degradation under arbitrary loading.

4. Shear wall design

The accuracy of the equivalent shell-wire model, including damage evolution, cumulative damage, strength and stiffness degradation, and computational cost were investigated under seismic loading. To this end, a shear wall structure was designed and analyzed under mainshock-aftershock seismic sequences. The building was a six-story reinforced structure with a height of 18 m that was intended for residential use in an area with a high seismic hazard level. The shear walls were designed in accordance with the simple tension and compression boundary element method (Simple T & C) [25, 34–36].

The pier of the wall was designed for a factored axial force (P_{f-top}) and moment (M_{f-top}) based on the loading combinations. The applied moment and axial force can be converted to the tantamount force applied to the bottom of the boundary elements Eqs. (9)–(10) as:

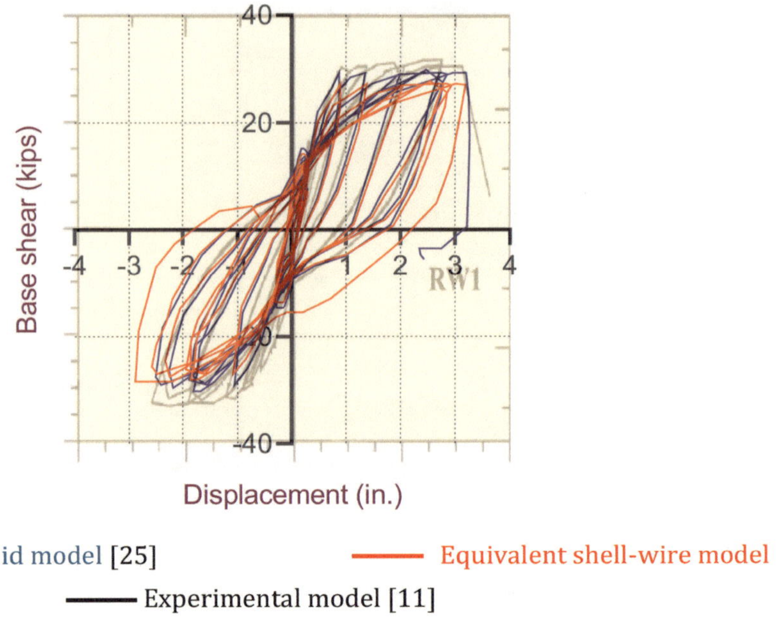

Figure 6.
Hysteresis curve comparison. 3D solid model [25]. Equivalent shell-wire model. Experimental model [11].

$$P_{left-top} = \frac{P_{f-top}}{2} + \frac{M_{f-top}}{(L_p - 0.5B_{1-left} - 0.5B_{1-right})} \tag{9}$$

$$P_{right-top} = \frac{P_{f-top}}{2} - \frac{M_{f-top}}{(L_p - 0.5B_{1-left} - 0.5B_{1-right})} \tag{10}$$

where L_p is the wall length and B_{1-left} and $B_{1-right}$ are the width of the left and right boundary members, respectively.

Under tensile or compression, the area required for rebars can be calculated using Eqs. (11) and (12), respectively, as:

$$A_{st} = \frac{P}{\phi_s f_y} \tag{11}$$

$$A_{sc} = \frac{\frac{Abs(P)}{P_{\max factor}} - \phi_c f'_c A_g}{\phi_s f_y - \phi_c f'_c} \tag{12}$$

where P is either $P_{left-top}$ or $P_{right-top}$, A_{st} is the area required for the steel reinforcements in the concrete design, ϕ_s is the steel resistance factor, f_y is the steel yield strength, $P_{\max factor}$ is defined by the shear wall design preferences (default is 0.80), ϕ_c is the concrete resistance factor for compression, f'_c is the concrete compressive strength, A_g is the gross area and is equal to the total area (ignoring reinforcements), and A_{sc} is the area required for steel reinforcements in the concrete design.

It should be noted that for negative values of A_{sc}, no compressive reinforcement is required; however, for all cases, the reinforcing ratio should be less than the maximum allowable ratio. **Table 1** presents the details of the shear walls, boundary members, and equivalent shell-wire model scheme of the six-story structure.

Story No.	Columns (boundary members)	Shear wall and boundary member	Equivalent shell-wire shear wall model
1 & 2	C50 50–16#6	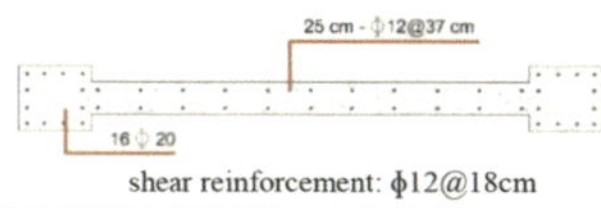	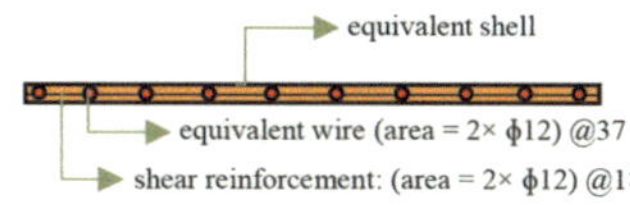
3 & 4	C50 50–16#6	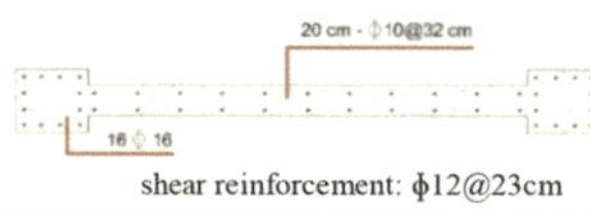	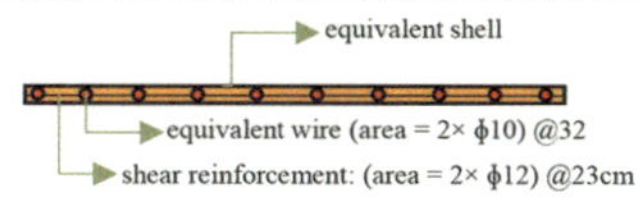
5 & 6	C50 50–16#4	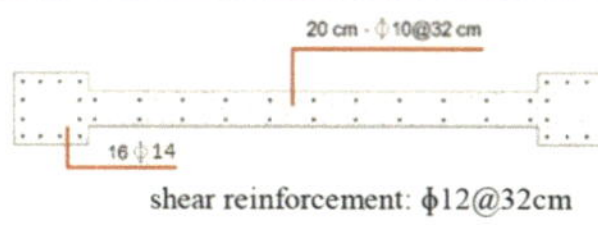	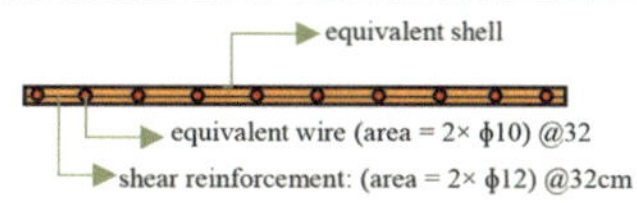

Table 1.
Details of shear walls, boundary members, and equivalent shell-wire model scheme.

5. Seismic sequence selection

The seismic behavior of the shear wall structure in Section 4 was tested under seven real mainshock-aftershock seismic sequences. It is known that near-field and far-field ground motions have different effect on the seismic behavior of a structure [37]; thus, near-field mainshock-aftershock records were selected from the PEER database [38]. The seismic declustering method (independent-dependent) was used

No.	Event	Station	Date	Magnitude (M_W)	PGA (g)
1	Friuli-Italy (Mainshock)	Tolmezzo	1976/05/06	6.5	0.35
	Friuli-Italy (Aftershock)		1976/05/07	5.2	0.11
2	Cape Mendocino (Mainshock)	Petrolia	1992/04/25	7.1	0.66
	Cape Mendocino (Aftershock)		1992/04/26	6.6	0.49
3	Mammoth Lakes (Mainshock)	Convict Creek	1980/05/25	6.6	0.42
	Mammoth Lakes (Aftershock)		1980/05/25	5.7	0.37
4	Imperial Valley (Mainshock)	El Centro Array #4	1979/10/15	6.5	0.32
	Imperial Valley (Aftershock)		1979/10/15	5.0	0.23
5	Northridge (Mainshock)	Newhall-Fire Station	1994/01/17	6.6	0.58
	Northridge (Aftershock)		1994/01/18	5.2	0.20
6	Northridge (Mainshock)	Rinaldi Receiving Station	1994/01/17	6.6	0.86
	Northridge (Aftershock)		1994/01/17	5.2	0.52
7	Whittier Narrows (Mainshock)	LA-Obregon Park	1987/10/01	5.9	0.42
	Whittier Narrows (Aftershock)		1987/10/01	5.2	0.34

Table 2.
Specifications of mainshock-aftershock records [25].

to choose the seismic sequences. In this method, the sequences were empirically determined using the data sequences from previous earthquakes and by measurement of the space–time history [39]. The magnitude of the mainshock should be greater than that of the aftershock and there are no limitations for peak ground acceleration (PGA) but the ratio of aftershock PGA to mainshock PGA of the sequences was considered to be less than 1 for this study [25, 26].

The shear wave velocity of the ground motion records agreed with the seismic design assumptions ($375 < V_s$ (m/s) < 750). The specifications of the mainshock-aftershock records are listed in **Table 2**.

Note that an appropriate zero acceleration interval was considered between the mainshock and aftershock to prevent the undesirable vibrations [40].

6. Analytical results

6.1 Inter-story drift ratio

The inter-story drift ratio is a fundamental structural parameter that can be used for the study of seismic behavior of structural [41]. It presents valuable data that can be used for structural judgment and analysis. The use of seismic sequences made it possible to compare the inter-story drift ratio of the mainshock and mainshock-aftershock for both the 3D solid and equivalent shell-wire models of the six-story structure. The behavior of the 3D solid model under seismic sequences was previously investigated [25].

The use of seismic sequences made the cumulative damage easily recognizable during this process. **Figure 7** shows the drift ratio for both the 3D solid and equivalent shell-wire models for both the mainshock and mainshock-aftershock. This section examined the accuracy of the equivalent shell-wire model under the uncertainty of cyclic loading (seven real mainshock-aftershock) while considering the cumulative damage, damage evolution, strength, and stiffness degradation.

As expected, the seismic sequences caused plastic strain as well as an increase in the drift ratio (cumulative damage), chiefly in story one [25, 26, 42]. Story one experienced the highest drift ratio and plastic strain and showed the greatest difference between the 3D solid model and equivalent shell-wire model results. On average, there was a difference of about 2% and 3% in the inter-story drift ratio between the 3D solid model and equivalent shell-wire models under the mainshock and mainshock-aftershock, respectively.

Note that the size of the mesh elements was the same for both models.

6.2 Changes in inter-story drift ratio (range of inter-story drift ratio)

The previous section (Section 6.1) showed that there is a good agreement between the inter-story drift ratios of the 3D solid model and the equivalent shell-wire model (under a 3% difference). It can be seen from **Figure 7** that higher plastic strain (high inter-story drift ratio) increases the difference between the 3D solid model and the equivalent shell-wire model. Also, seismic sequences caused plastic strain growth (inter-story drift ratio growth) in some stories. So, it is possible to define the range of inter-story drift ratio that shows changes in drift ratio under mainshock (lower range) and mainshock-aftershock (higher range). This parameter would be valuable to

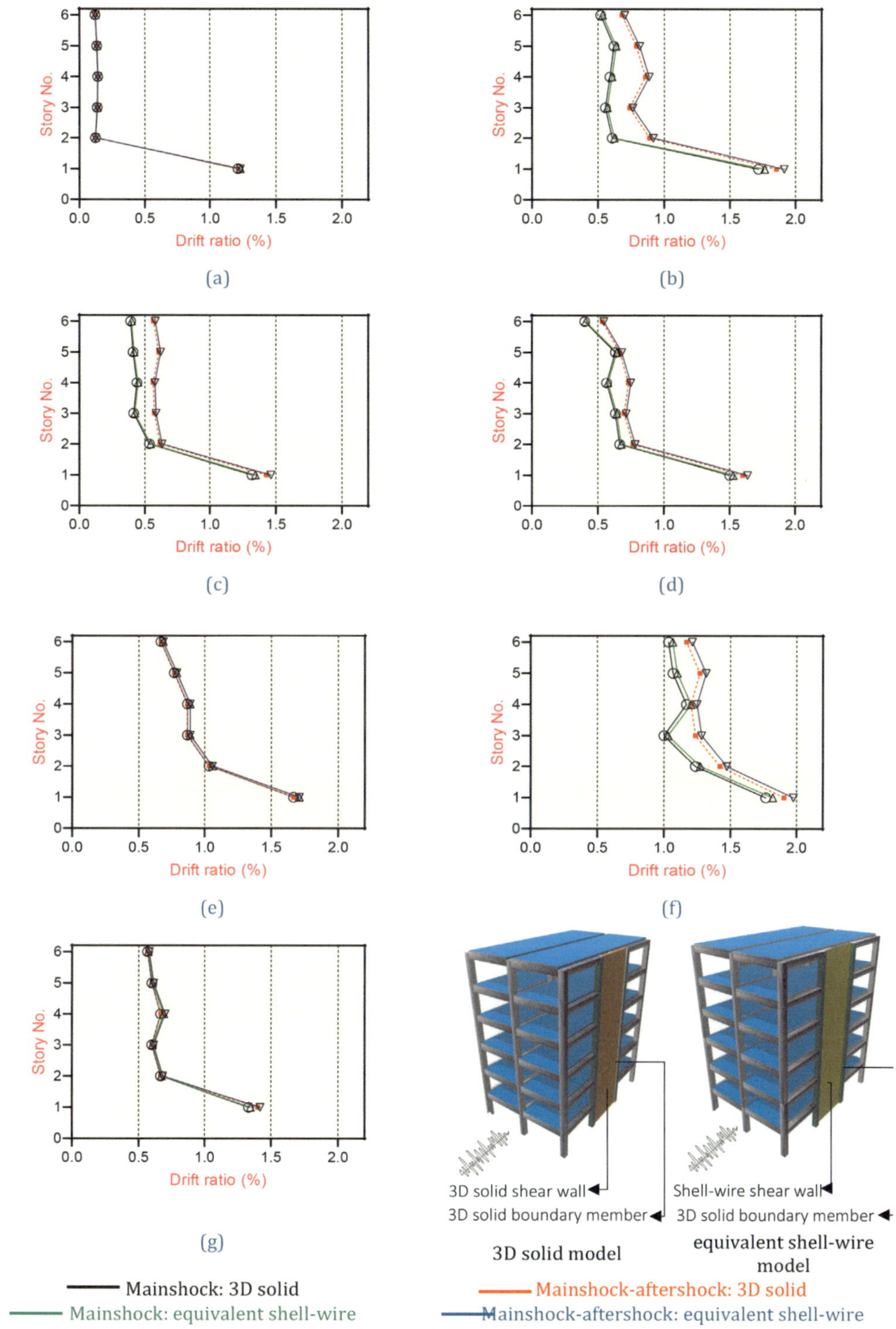

Figure 7.
Comparison of inter-story drift ratio: (a) Friuli; (b) Cape Mendocino; (c) Mammoth Lakes; (d) Imperial Valley; (e) Northridge (Newhall-Fire Station); (f) Northridge (Rinaldi Receiving Station); (g) Whittier narrows.

examine the accuracy of the proposed equivalent shell-wire model under high plastic strain. In other words, the equivalent shell-wire model can consider accurately cumulative damage, strength, and stiffness degradation if there is a good agreement between the range of inter-story drift ratio (drift ratio of mainshock and mainshock-aftershock) of the 3D solid model and the equivalent shell-wire model.

Figure 7 shows that the calculated inter-story drift ratios for both the mainshock and mainshock-aftershock obtained from the 3D solid and equivalent shell-wire models were almost similar. The inter-story drift ratios changed (range of the inter-story drift ratio) under the mainshocks and mainshocks-aftershocks for the models, as can be seen in **Figure 8**. Story one recorded the highest inter-story drift ratio, plastic strain, and damage under the mainshocks and mainshocks-aftershocks; thus, this story was selected for **Figure 8**.

Figure 8 shows that, as the plastic strain and inter-story drift ratio increased, the difference between the results of models increased. For example, the ranges of the drift ratios for the Friuli mainshock and mainshock-aftershock for both models were similar (roughly 1% difference). However, for the Cape Mendocino mainshock-aftershock, there was about a 3.14% difference. **Table 3** lists the differences in the mainshock-aftershock inter-story drift ratios of story one. As seen, there was a less than 4% difference between the 3D solid and the equivalent shell-wire model in all

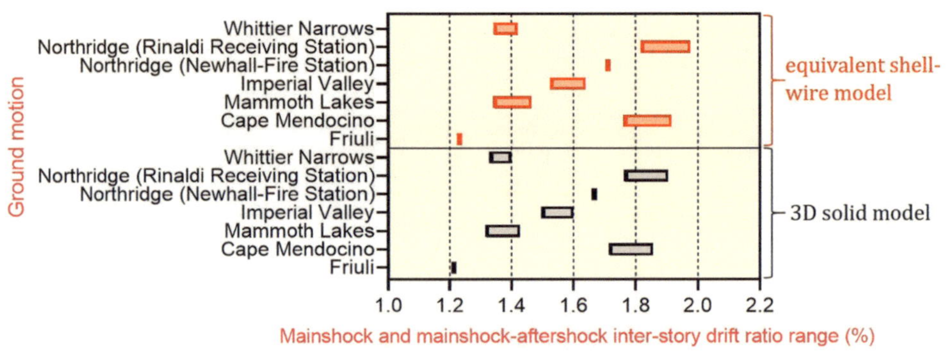

Figure 8.
Range of story one inter-story drift ratios for mainshock and mainshock-aftershock.

1	2	3	4
Mainshock-aftershock	3D solid model (%)	Equivalent shell-wire model (%)	Difference bet. 3 and 2 (%)
Friuli	1.21	1.23	1.47
Cape Mendocino	1.85	1.91	3.14
Mammoth Lakes	1.42	1.46	2.43
Imperial Valley	1.60	1.64	2.44
Northridge (Newhall-Fire Station)	1.66	1.71	2.53
Northridge (Rinaldi Receiving Station)	1.90	1.97	3.66
Whittier Narrows	1.40	1.42	1.40

Table 3.
Inter-story drift ratio comparison of the story one.

cases, which is not remarkable (2.44% on average). Thus, the equivalent shell-wire model provided the appropriate results.

6.3 Computational cost

The computational cost could be defended as "the amount of time required to complete certain operation" [43]. In FE analysis, the computational cost is the execution time per increment or the total time required to analyze a job (sum of execution time for all increments). This study tried to compare the computational cost of the 3D solid model and the equivalent shell-wire model under certain operations (under mainshock-aftershocks). Thus, the total time required to analyze the 3D solid model and equivalent shell-wire model under each seismic sequence was considered as the computational cost in the current study. For each seismic sequence, the computational costs of the 3D solid model and equivalent shell-wire model were compared.

As stated, because of the complexity of 3D solid models and their high computational cost, studies have generally focused on SDOF systems or 2D frames having low heights. Similar 3D structures have been neglected. It is also known that structures can show different seismic behaviors under seismic sequences depending on the 2D or 3D modeling and that 3D models are preferred [12]. Sections 3, 6.1, and 6.2 demonstrated that the equivalent shell-wire model was able to provide analytical results that were similar to those of 3D solid complex models that had high computational costs. Equivalent shell-wire and 3D solid models also showed similar results for cumulative damage, damage evolution, strength, and stiffness degradation. This makes it necessary to determine the relative computational costs of the equivalent shell-wire and 3D solid models.

Table 4 lists and compares the time required for analysis using the equivalent shell-wire and 3D solid models. As it is known that the computational cost is equivalent to the time required for analysis, on average, the equivalent shell-wire model had a 39% lower computational cost than the 3D solid model, which is remarkable. As the inter-story drift ratio or plastic strain increased, the computational cost efficiency of the equivalent shell-wire model over 3D solid model increased. It can be said that, considering the accuracy and computational cost, the equivalent shell-wire model performed better at a less computational cost. This approach also can be used to study complex and taller structures and even use finer mesh elements to increase accuracy.

Mainshock-aftershock	Comparison of time required for analyses (%)
Friuli	−38%
Cape Mendocino	−40%
Mammoth Lakes	−39%
Imperial Valley	−39%
Northridge (Newhall-Fire Station)	−39%
Northridge (Rinaldi Receiving Station)	−40%
Whittier Narrows	−39%

Table 4.
Computational cost comparison.

7. Conclusions

The realistic seismic behavior of structures, chiefly shear wall structures, under seismic sequences is of major concern in seismic engineering. In general, 3D solid structural models show more accurate results than 2D models; however, because of cumulative damage, damage evolution, and strength and stiffness degradation, analysis of 3D solid models under seismic sequences accrue higher computational costs, which means that previous studies have generally only investigated 2D models or low-height 3D solid models. The use of an equivalent shell-wire model is an appropriate technique for generating accurate results while decreasing the computational cost for shear wall structures. The results of this study could be summarized as follows:

- Comparison of the area under the capacity curve of a shear wall modeled using the 3D solid model and equivalent shell-wire model, which is equal to absorbed energy, showed that there was only a slight difference between the curves. The displacement-based shear curves showed less than a 5% difference; however, compared with the 3D solid model, the equivalent shear-wire model required only about 50% of the computational cost.

- The cyclic behavior of the RW1 shear wall modeled using the equivalent shell-wire model was compared with the results of an experimental study. The generated hysteresis curves showed good agreement between the equivalent shell-wire model and the experimental model, including for cyclic strength and stiffness degradation.

- In order to assess the accuracy of the equivalent shell-wire model, including the cumulative damage, damage evolution, strength, and stiffness degradation, a six-story shear wall structure was designed and modeled using the 3D solid model and equivalent shell-wire model in Abaqus. As expected, the seismic sequences increased the drift ratio and damage in the intended structure. Comparison of the inter-story drift ratios under the mainshock and mainshock-aftershock showed good agreement between the 3D solid model and equivalent shell-wire models. For example, in story one, on average, there was about a 1.5% and 2.5% difference in the inter-story drift ratio between the 3D solid and equivalent shell-wire model under mainshock and mainshock-aftershock, respectively.

- Assessment of the inter-story drift ratio showed that an increase in the plastic strain or damage increased the difference between the results from the 3D solid and equivalent shell-wire wire models. For example, under the Friuli seismic sequence, there was about a 1% difference in the inter-story drift ratio between the model for story one. This difference was 3.14% under the Cape Mendocino seismic sequences, which could be because of the higher drift ratio (plastic strain) experienced by story one under the Cape Mendocino mainshock-aftershock (1.91%) compared to the Friuli mainshock-aftershock (1.23%).

- Comparison of the story one inter-story drift ratio for the 3D solid and equivalent shell-wire models showed less than a 4% difference between models for all sequences used (almost 2.5% on average). This indicates that the equivalent shell-wire model provided accurate results and could appropriately consider the cumulative damage, damage evolution, strength, and stiffness degradation.

- Along with the accurate and appropriate results of the equivalent shell-wire model, on average, this model required 39% less computational cost than the 3D solid model. This suggests that it is possible to analyze structural models of greater height, more complexity, or with finer mesh to increase accuracy at the same computational cost and accuracy of a 3D solid model.

Acknowledgements

There is no acknowledgment.

Conflict of interest

The authors declare no conflict of interest.

Notes/thanks/other declarations

There are no notes/thanks/other declarations.

Author details

Omid Karimzade Soureshjani and Ali Massumi*
Kharazmi University, Tehran, Iran

*Address all correspondence to: massumi@khu.ac.ir

References

[1] Galal K, El-Sokkary H. Advancement in modeling of RC shear walls. In: The 14th World Conference on Earthquake Engineering. Beijing: China; 2008

[2] Astaneh-Asl A. Seismic Behavior and Design of Composite Steel Plate Shear Walls. Moraga, CA: Structural Steel Educational Council; 2002

[3] Qu B, Bruneau M. Design of steel plate shear walls considering boundary frame moment resisting action. Journal of Structural Engineering. 2009;**135**(12): 1511-1521. DOI: 10.1061/(ASCE) ST.1943-541X.0000069

[4] White MW, Dolan JD. Nonlinear shear-wall analysis. Journal of Structural Engineering. 1995;**121**(11):1629-1635. DOI: 10.1061/(ASCE)0733-9445(1995) 121:11(1629)

[5] Amadio C, Fragiacomo M, Rajgelj S. The effects of repeated earthquake ground motions on the non-linear response of SDOF systems. Earthquake Engineering & Structural Dynamics. 2003;**32**(2):291-308. DOI: 10.1002/ eqe.225

[6] Wallace JW. New methodology for seismic design of RC shear walls. Journal of Structural Engineering. 1994;**120**(3): 863-884. DOI: 10.1061/(ASCE) 0733-9445(1994)120:3(863)

[7] Massumi A, Karimi N, Ahmadi M. Effects of openings geometry and relative area on seismic performance of steel shear walls. Steel and Composite Structures. 2018;**28**(5):617-628. DOI: 10.12989/scs.2018.28.5.617

[8] Tarigan J, Manggala J, Sitorus T. The effect of shear wall location in resisting earthquake. In IOP Conference Series: Materials Science and Engineering. Vol. 309, No. 1. Sumatera Utara, Indonesia: IOP Publishing; 1 Feb 2018. p. 012077

[9] Khouri MF. Drift limitations in a shear wall considering a cracked section. International Journal of Reliability and Safety of Engineering Systems and Structures. 2011;**1**(1):31-38

[10] Graves A. Adaptive computation time for recurrent neural networks. arXiv preprint arXiv:1603.08983. 29 Mar 2016;**Ver. 6**:1-19. DOI: 10.48550/ arXiv.1603.08983

[11] Thomsen JH IV, Wallace JW. Displacement-based design of slender reinforced concrete structural walls—Experimental verification. Journal of Structural Engineering. 2004;**130**(4): 618-630. DOI: 10.1061/(ASCE) 0733-9445(2004)130:4(618)

[12] Ruiz-García J, Yaghmaei-Sabegh S, Bojórquez E. Three-dimensional response of steel moment-resisting buildings under seismic sequences. Engineering Structures. 2018;**15**(175):399-414. DOI: 10.1016/j.engstruct.2018.08.050

[13] Hatzivassiliou M, Hatzigeorgiou GD. Seismic sequence effects on three-dimensional reinforced concrete buildings. Soil Dynamics and Earthquake Engineering. 2015;**1**(72):77-88. DOI: 10.1016/j.soildyn.2015.02.005

[14] Ruiz-García J, Negrete-Manriquez JC. Evaluation of drift demands in existing steel frames under as-recorded far-field and near-fault mainshock–aftershock seismic sequences. Engineering Structures. 2011;**33**(2):621-634. DOI: 10.1016/j.engstruct.2010.11.021

[15] Calladine CR. Theory of Shell Structures. New Rochelle, New York: Cambridge University Press; 16 Feb 1989

[16] Systemes D. Getting, Start with ABAQUS Tutorial, ABAQUS Version 6.14. Providence, RI: Dassault Systèmes; 2017

[17] Johnson S. Comparison of Nonlinear Finite Element Modeling Tools for Structural Concrete. CEE561 Project. Champaign, IL, USA: University of Illinois; 2006

[18] Systèmes D. Getting Start with ABAQUS Tutorial. ABAQUS Version 2020. Rhode Island, USA: Systèmes D; 2020

[19] Tsaparli V, Kontoe S, Taborda DM, Potts DM. The importance of accurate time-integration in the numerical modelling of P-wave propagation. Computers and Geotechnics. 2017;**1**(86): 203-208. DOI: 10.1016/j.compgeo.2017.01.017

[20] Hilber HM, Hughes TJ, Taylor RL. Improved numerical dissipation for time integration algorithms in structural dynamics. Earthquake Engineering & Structural Dynamics. 1977;**5**(3):283-292. DOI: 10.1002/eqe.4290050306

[21] Hibbitt HD, Karlsson BI. Analysis of Pipe Whip. Providence, USA: Hibbitt and Karlsson Inc; 1979

[22] Weaver W, Johnston PR. Structural Dynamics by Finite Elements. Englewood Cliffs, NJ: Prentice-Hall; 1987

[23] Bathe KJ. Finite Element Method. Hoboken, New Jersey: Wiley Encyclopedia of Computer Science and Engineering; 15 Mar 2007. pp. 1-2. DOI: 10.1002/9780470050118.ecse159

[24] Náprstek J, Horáček J, Okrouhlík M, Marvalová B, Verhulst F, Sawicki JT, editors. Vibration Problems ICOVP 2011: The 10th International Conference on Vibration Problems. Prague, Czech Republic: Springer Science & Business Media; 14 Dec 2011

[25] Soureshjani OK, Massumi A. Seismic behavior of RC moment resisting structures with concrete shear wall under mainshock–aftershock seismic sequences. Bulletin of Earthquake Engineering. 2022;**20**(2):1087-1114. DOI: 10.1007/s10518-021-01291-x

[26] Soureshjani OK, Nouri G. Seismic assessment and rehabilitation of a RC structure under mainshock-aftershock seismic sequences using beam-column bonded CFRP strategy. Bulletin of Earthquake Engineering. 2022;**29**:1-28. DOI: 10.1007/s10518-022-01382-3

[27] Johnson GR, Cook WH. Fracture characteristics of three metals subjected to various strains, strain rates, temperatures and pressures. Engineering Fracture Mechanics. 1985; **21**(1):31-48. DOI: 10.1016/0013-7944(85)90052-9

[28] Lubliner J, Oliver J, Oller S, Oñate E. A plastic-damage model for concrete. International Journal of Solids and Structures. 1989;**25**(3):299-326. DOI: 10.1016/0020-7683(89)90050-4

[29] Hillerborg A, Modéer M, Petersson PE. Analysis of crack formation and crack growth in concrete by means of fracture mechanics and finite elements. Cement and Concrete Research. 1976;**6**(6):773-781. DOI: 10.1016/0008-8846(76)90007-7

[30] Lee J, Fenves GL. Plastic-damage model for cyclic loading of concrete structures. Journal of engineering mechanics. 1998;**124**(8):892-900. DOI: 10.1061/(ASCE)0733-9399(1998)124:8(892)

[31] Wang B, Huo G, Sun Y, Zheng S. Hysteretic behavior of steel reinforced

concrete columns based on damage analysis. Applied Sciences. 2019;**9**(4): 687. DOI: 10.3390/app9040687

[32] Londoño JM, Neild SA, Cooper JE. Identification of backbone curves of nonlinear systems from resonance decay responses. Journal of Sound and Vibration. 2015;**21**(348):224-238. DOI: 10.1016/j.jsv.2015.03.015

[33] Sutoyo D. Hysteretic Characteristics of Wood-Frame Structures under Seismic Motions. USA: California Institute of Technology; 2009

[34] Design of a Building Using: ETABS Step by Step Procedure of Designing Building in ETABS. Etabs Tutorial. Berkeley, California, USA: Etabs Computers and Structures Inc.; 2010

[35] Computers & Structures Inc. Shear Wall Design Manual ACI 318–14. Berkeley, California, USA: Computers & Structures Inc.; 2016

[36] Kaveh A, Zakian P. Optimal seismic design of reinforced concrete shear wall-frame structures. KSCE Journal of Civil Engineering. 2014;**18**(7):2181-2190. DOI: 10.1007/s12205-014-0640-x

[37] Soureshjani OK, Massumi A. Effect of near and far-field earthquakes on RC bridge with and without damper. Earthquakes and Structures. **17**:533-543. DOI: 10.12989/eas.2019.17.6.000

[38] PEER Ground Motion Database. Berkeley: California: Pacific Earthquake Engineering Research Centre, University of California. Available from: https://ngawest2.berkeley.edu/

[39] Hainzl S, Steacy D, Marsan S. Seismicity models based on Coulomb stress calculations. Community Online Resource for Statistical Seismicity Analysis. 2010;**10**(1):1-26. DOI: 10.5078/corssa-32035809. Available from: http://www.corssa.org

[40] Pirooz RM, Habashi S, Massumi A. Required time gap between mainshock and aftershock for dynamic analysis of structures. Bulletin of Earthquake Engineering. 2021;**19**(6):2643-2670. DOI: 10.1007/s10518-021-01087-z

[41] Yang D, Pan J, Li G. Interstory drift ratio of building structures subjected to near-fault ground motions based on generalized drift spectral analysis. Soil Dynamics and Earthquake Engineering. 2010;**30**(11):1182-1197. DOI: 10.1016/j.soildyn.2010.04.026

[42] Massumi A, Sadeghi K, Ghaedi H. The effects of mainshock-aftershock in successive earthquakes on the response of RC moment-resisting frames considering the influence of the vertical seismic component. Ain Shams Engineering Journal. 2021;**12**(1):393-405. DOI: 10.1016/j.asej.2020.04.005

[43] Burgos DA, Vejar MA, Pozo F, editors. Pattern Recognition Applications in Engineering. Pennsylvania, USA: IGI Global; 27 Dec 2019

Section 2

Underground Construction

Chapter 5

Design Method and Construction Technology in Tunnel Engineering under Complex Geological Conditions

Siming Tian, Yang Xue, Zhenyan Tian, Shaoshuai Shi, Weidong Guo, Wei Wang and Ruijie Zhao

Abstract

Over the past 40 years of reform and opening up, China's railway tunnel construction industry has made great progress. The completed railway tunnels account for nearly 90% of the total length of China's railway tunnels. At present, China's tunnel construction technology is relatively advanced, but with the development of tunnel engineering in the southwest, railway tunnel construction is facing a series of major and complex geological problems. This chapter mainly introduces the tunnel design theory and method, support structure system, construction technology, and equipment under complex geological conditions. In addition to the technical problems and challenges faced by the current railway tunnel engineering, this chapter puts forward the development of active support collaborative control concept and technology, digital survey and design, intelligent construction, and intelligent operation and maintenance, so as to provide a useful reference for promoting the development of tunnel engineering.

Keywords: tunnel engineering, complex geology, design method, construction technology, supporting structure

1. Introduction

In the twenty-first century, China's railway tunnel engineering has developed rapidly. Especially with the construction of high-speed railways and intercity railways and the implementation of the national "western development" strategy, railway tunnels show that the requirements due to high construction standards and complex geology, and complex special tunnels are gradually increasing, The scale of urban tunnels in complex environment and underwater tunnels crossing rivers and seas has increased rapidly. The completion and opening of these tunnel projects have made outstanding contributions to traffic and economic development. At the same time, there has been an accumulation of a large number of scientific and technological achievements and practical experience in railway tunnel engineering.

After 40 years of development, China's railway tunnel has made great progress in basic theory, design method, construction technology, and equipment development. It has made major achievements and breakthroughs in high-speed railway tunnel, large-section loess tunnel, underwater tunnel, and TBM tunnel in complex and dangerous mountainous areas. It has significantly improved the technical level of railway tunnel construction and demonstrated the international influence of China's railway tunnel construction [1]. However, with the advancement of national major railway construction and the construction of underwater tunnels across rivers and sea, China's railway tunnels are also facing technical challenges in structural durability, the construction of large buried deep and ultra-long railway tunnels in high sea uplift seismic areas, large-diameter shield high water pressure long-distance construction, and safe operation of tunnel engineering. Therefore, combined with the current needs of tunnel engineering construction, this chapter summarizes and analyzes the tunnel engineering design methods and construction technology under complex geological conditions, and puts forward the direction of development and prospect according to the technical problems and challenges faced by the current railway tunnel engineering.

2. Theory and method of railway tunnel design

Since the 1980s, China has always adopted the standard design and analogy design methods based on experience. In recent years, China's railway design has incorporated the concepts of "New Austrian Tunneling Method [2]," "Norwegian method [3]," and "New Italian method [4]." At the same time, a semi-quantitative approach to railway tunnel design has been developed, which incorporates information about the complex and diverse terrains, and the geological and climatic conditions affecting China's railway tunnels. Structural design has changed from allowable stress method to probability reliability method based on limit state, and numerical simulation methods based on finite element and finite difference have also begun to be applied to structural design. Tunnel design quality has been greatly improved, and tunnel design ideas have also undergone major changes. There is a shift in design from one that relies solely on the lining bearing to those that rely on the surrounding rock. Many studies have been carried out in the aspects of surrounding rock stability evaluation and classification, and surrounding rock deformation control design. Finally, a tunnel construction method with Chinese characteristics based on active control of surrounding rock deformation was formed, which greatly promoted the development of railway tunnel design theory and methods [1].

2.1 Design method based on surrounding rock stability evaluation and classification

The standard design and analogy design of railway tunnel in China are mainly based on the evaluation and classification of surrounding rock stability. The stability of surrounding rock is directly related to the safety of tunnel construction and is the core issue of design and construction. After a lot of engineering practice and analysis, it is proposed to take the self-stability of surrounding rock as the index and formulate a unified surrounding rock classification standard, as shown in **Table 1**, which has guided the design and construction practice of railway tunnel for a long time.

However, in the actual construction process, the surrounding rock revealed by excavation does not correspond to the basic level of surrounding rock, so it is difficult to judge on-site, and it is easy to have weak or strong support parameters. In recent years, railway builders have carried out a lot of scientific research and test work to improve and strengthen the quantitative classification of surrounding rock. By refining the basic quality index of surrounding rock (BQ) and the elastic wave velocity range of surrounding rock with different lithologies, the concept of surrounding rock sub-grade is proposed, and the surrounding rock of grades III, IV, and V is divided into two sub-grades, respectively. The index combination corresponding to each sub-grade is the sub-grade division standard. Finally, the BQ value range corresponding to each sub-grade surrounding rock is determined, which well guides the design and construction of railway tunnel. The sub-grade and stability of railway tunnel surrounding rock are shown in **Table 1**.

The surrounding rock classification method of railway tunnel has been relatively perfect. The measurement while drilling technology such as advance drilling parameters has developed rapidly, and the intelligent construction technology of tunnel has been continuously promoted. It would be possible to establish a method to map rock mass quality indicators based on drilling parameters, and then dynamically and automatically quantify and interpret the grade of surrounding rock.

Surrounding rock level		**Stability description of surrounding rock at all levels**
Basic level	**Subclass**	
I	—	The surrounding rock is stable, no collapse, and rockburst may occur
II	—	If the exposure time is long, local small collapses may occur, and the damage is mainly due to falling blocks, the side walls are stable, and the roof of the gentle rock layer with poor interlayer bonding is easy to collapse; the temporary self-stabilizing span is 17–20 m
III	III 1	When the arch is not supported, a small collapse can occur, the side wall is basically stable, and the blasting vibration is too large and easy to collapse; the temporary self-stabilizing span is 14 $\sim$ 16 m
	III 2	When the arch is not supported, a small collapse can occur, the side wall is basically stable, and the blasting vibration is too large and easy to collapse; the temporary self-stabilizing span is 10–13 m
IV	IV 1	When the arch is unsupported, it can collapse greatly, and the side wall sometimes loses its stability; the temporary self-stabilizing span is 7–9 m
	IV 2	When the arch is not supported, it can cause a large collapse, and the side wall sometimes loses its stability; the temporary self-stabilizing span is 5–6 m
V	V 1	The surrounding rock is easy to collapse, and if it is not handled properly, there will be large collapse, the side wall often has small collapse, and the surface is prone to subsidence or collapse to the surface during shallow burial; the temporary self-stabilizing span is 3–4 m
	V 2	The surrounding rock is easy to collapse, and if it is not handled properly, there will be large collapse, the side wall often has small collapse, and the surface is prone to subsidence or collapse to the surface during shallow burial; the temporary self-stabilizing span is less than 3 m
VI	—	The surrounding rock is easy to collapse and deform. When there is water, the soil and sand often gushing out together with the water. When buried shallowly, it is easy to collapse and penetrate to the surface; there is no self-stability

Table 1.
Sub-classification and stability of surrounding rock of railway tunnel [5].

2.2 Design method for deformation control of surrounding rock

The research team of high-speed railway tunnel surrounding rock stability control technology led by Zhao Yong and others [1] innovatively proposed a set of tunnel construction technology methods with Chinese characteristics. Its principle is that on the premise of ensuring the stability of surrounding rock, small tremie advance grouting, reasonable lining support, and other means are used to control the deformation of surrounding rock, give play to the coordination between the support structure and the surrounding rock structure, make full use of the self-supporting capacity of surrounding rock, and form a fast and durable tunnel stability structure system. As shown in **Figure 1**, advance reinforcement is used to prevent surrounding rock collapse and control the scope and extent of damage to the excavation surface. The initial support structure bears all additional stresses released by the surrounding rock due to excavation to help the surrounding rock share the load. As a safety reserve, the secondary lining shares part of the load with the primary support.

This method considers that the surrounding rock of the tunnel is composed of "shallow surrounding rock" and "deep surrounding rock." The surrounding rock in the loose area needs to be supported in time for the load of this part of the surrounding rock. Outside this range, the surrounding rock with good overall stability and capable of bearing the stratum load is called "deep surrounding rock." It is composed of "loading layers," which start from the shallow surrounding rock interface and appear alternately [1].

2.3 Design method of mechanized large section of tunnel

The tunnel construction team of the Hubei section of the Zhengwan High-speed Railway has carried out scientific research and field practice related to the construction of large-scale mechanized large-section tunnels. In summary, the relevant methods of railway tunnel mechanized large-section design and construction are explained as follows [6, 7].

In terms of mechanized supporting design, it covers four operation areas: advance support, excavation, initial support, and secondary lining, which are divided into basic supporting and reinforced supporting according to the perfection of the configured machinery. Advance support is one of the auxiliary measures to ensure the

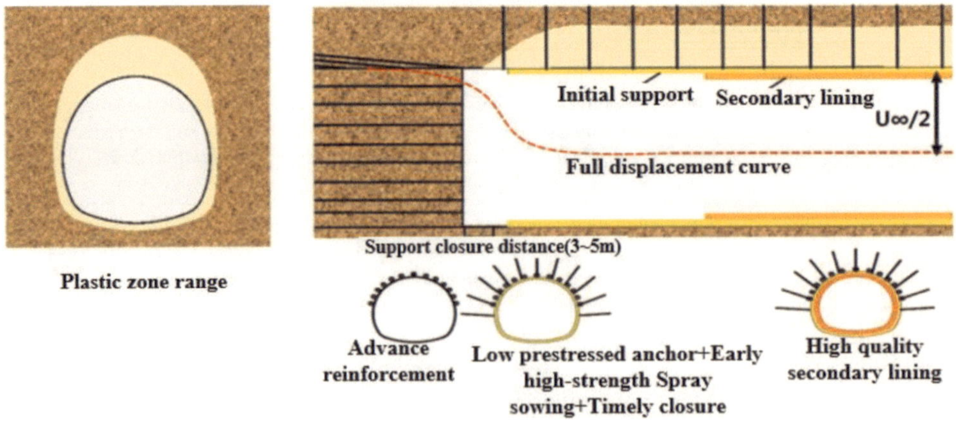

Figure 1.
Schematic diagram of the control design principle of the full deformation process of the tunnel.

stability of the excavation face of the tunnel, which is taken ahead of the excavation of the tunnel face. The main methods are pipe shed, small conduit, horizontal jet grouting pile, etc. Excavation refers to the work of loosening, breaking, excavating, and transporting soil or rock for mucking during the construction of tunnels. The initial support refers to the support carried out immediately after excavation, which generally includes shotcrete, shotcrete with anchor rod, shotcrete anchor rod, and steel frame combined support. Secondary lining is a formwork concrete or reinforced concrete lining applied inside the initial support during tunnel construction, and together with the initial support, it forms a composite lining.

In terms of construction method design, there are full-section methods and microstep methods. In the aspect of face stability evaluation, a combination of qualitative evaluation and quantitative evaluation is adopted, which is divided into three types: stable, temporary stable, and unstable.

In terms of advance support design, advance support measures (including surface shotcrete, advance small conduits, pipe sheds, surface anchors, advance grouting, etc.) are determined according to the results of the face stability evaluation. The engineering analogy method and the limit equilibrium method are used to analyze and determine the parameters. The design model of the tunnel face advanced support is shown in **Figure 2**, and the calculation formula is shown in formula (1) [8].

In terms of the design method of tunnel support, the load structure model is used for calculation. The surrounding rock pressure in the shallow buried and eccentric pressure sections is adopted according to the formula value of the tunnel design code, and the deep buried section is determined according to the measured value of deformation and stress.

$$K = \frac{P_1 + \beta_1 \alpha_1 F_c}{\beta_2 \left(F_w + \alpha_2 F_q\right)} + \beta_3 \tag{1}$$

where β_1, β_2, β_3 are coefficients related to the internal friction angle of surrounding rock; α_1 is the vertical deformation pressure reduction coefficient of the disturbance

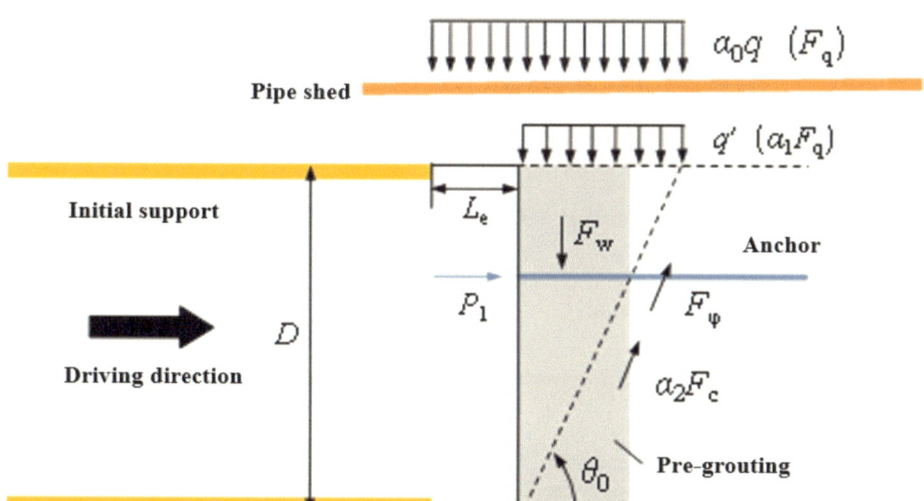

Figure 2.
Design model of advance support of the tunnel face. Note: L_e is the length of the unsupported section, m; θ_o is the failure angle of the tunnel face; D is the height of the tunnel, m.

section in front of the tunnel face under the pipe shed support; α_2 is the cohesion increase coefficient of surrounding rock after pre grouting reinforcement of tunnel face; F_c is the cohesion force of the sliding surface, kN; F_q is the resultant force of vertical deformation pressure of the sliding body, kN; F_w is the self-weight of the sliding body, kN; P_1 is the bolt support force of the tunnel face, N; K is the stability coefficient of the tunnel face.

This method has now formed the standard for China National Railway Group Co., Ltd. (draft for approval in 2021). The promulgation of this standard will greatly promote the mechanization level of railway tunnel drilling and blasting construction and provide technical support for the intelligent construction of railway tunnels.

2.4 Total safety factor method for tunnel support structure design

Xiao Mingqing and his team summarized and proposed the "total safety factor method for tunnel support structure design" through years of tunnel engineering design, construction, and research [9]. The total safety factor is a factor used to simultaneously evaluate the overall structural safety of initial lining and secondary lining. The technical idea of this method is to regard the interaction relationship between the support structure and the surrounding rock as the relationship between the action force and the reaction force. The deformation compatibility between supporting structure and surrounding rock is not strictly considered, which greatly simplifies the problem to be solved. The judgment of whether the tunnel needs support and the calculation of the support force need to be determined after numerical analysis using various constitutive models that are consistent with the actual situation. The intrinsic safety and deformation of the support structure are calculated by the load structure model to realize the safety evaluation and quantitative design of the support parameters [10]. **Figure 3** is a composite structure model formed by combining the three-layer structure of bolt-surrounding rock-bearing arch, shotcrete layer, and secondary lining. In it, the purpose of spray layer is to reinforce the tunnel structure and improve the stability and safety of the tunnel. The secondary lining is a molded concrete or reinforced concrete lining constructed on the inner side of the primary support, which forms a composite lining together with the primary support

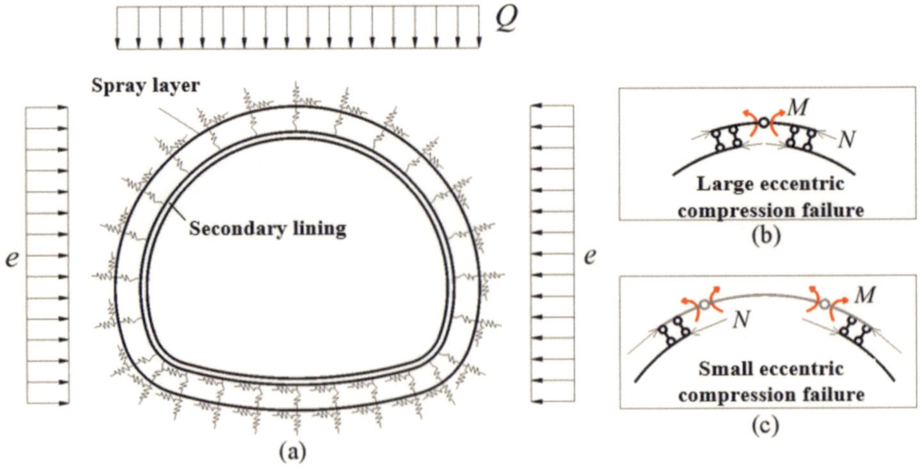

Figure 3.
Calculation model of total safety factor method.

to achieve the effect of reinforcement support. Therefore, we use the model shown in **Figure 3** to calculate the total safety factor and evaluate the safety of the tunnel support structure. The relevant calculation formula is as follows [11]:

Construction stage (without secondary lining):

$$K_c = \eta K_1 + K_2 \tag{2}$$

Operational stage:

When using durability anchors:

$$K_{op} = \eta K_1 + \xi K_2 + K_3 \tag{3}$$

When using non-durable bolts:

$$K_{op} = \xi K_2 + K_3 \tag{4}$$

where K_c and K_{op} are the total safety factor in the construction stage and operation stage, respectively; K_1, K_2, and K_3 are the safety factors of anchor-rock bearing arch, initial support, and secondary lining, respectively; η is the correction factor for the safety factor of the anchor rock bearing arch; ξ is the correction factor of the safety factor of initial support.

The total safety factor during the construction period is recommended as follows: when the sprayed layer is made of steel fiber reinforced concrete or with a steel frame, it is not less than 1.8, and when the sprayed layer is made of plain concrete, it is not less than 2.1. The recommended total safety factor during the operation period is: not less than 3.0 when the secondary lining is made of reinforced concrete, and not less than 3.6 when the secondary lining is made of plain concrete. A tunnel that satisfies the above-mentioned total safety factor is a safe and good tunnel. When the stress of a section of the spray layer or secondary lining composite structure is greater than the safety bearing capacity, it will reach the damage stage. At this time, if the offset distance of the axial force N is large and the steel bar is configured appropriately, large eccentric compression failure will occur. If the offset distance of the axial force N is small but the configuration of the steel bar is excessive, small eccentric compression failure will occur. The calculation model of large and small eccentric compression failure is shown in **Figure 3(b)** and **(c)**.

This design method can easily calculate the safety factor of the tunnel at different stages, and it is hoped that it will be further improved through field measurement and model test, which provides a convenient calculation method for the quantitative design of the tunnel.

3. Tunnel construction technology

3.1 Informatization design and construction technology

In the 40 years of tunnel development in China, we have constantly summarized the practical experience of railway tunnel construction informatization, taking safety and quality management as the main line, and reference to domestic and foreign engineering construction informatization experience. An information-based design and construction technology based on advanced geological prediction and monitoring measurement has been formed.

The traditional advanced detection methods mainly focus on advanced drilling, seismic emission, electromagnetics, electrical methods, etc. In recent years, with the development of drilling equipment, advanced exploration technology has also been developed rapidly. Dozens of meters to thousands of meters can be achieved. At the same time, some new ideas of advanced tunnel geological prediction technology have been proposed, including tunnel-induced polarization technology, tunnel nuclear magnetic resonance technology, single-hole directional radar technology, and seismic wave water exploration technology. It has played a beneficial and positive role in promoting the advancement of advanced geological forecasting technology and improving engineering service capabilities. The three-pole bathymetric induced polarization method proposed by Shandong University, which can quantitatively estimate the water volume of water bodies within 30 m in front of the driving face, and the full-space transient electromagnetic method, which can identify and locate the water bodies within 80 m in front of the driving face, are suitable for quantitative detection in adverse geological conditions [12]. Important progress has been made in the field, especially the feasible method for estimating the water content of aquifers, and the establishment of a comprehensive advanced detection technology and system [13, 14]. The above achievements have been successfully applied in many projects with complex geological conditions, such as the Chengdu-Lanzhou Railway, the Chengdu-Kunming Railway, and the Beijing-Zhangjiakou High-speed Railway. It fully reflects the rapid development of advanced geological forecasting technology, which provides important support for the safe and efficient construction of railway tunnels in China.

Tunnel site monitoring and measurement is a key link in tunnel construction and the basis for realizing information-based design and construction. The traditional monitoring technology mainly relies on personnel entering the construction site and using traditional measuring tools for manual operation measurement; in recent years, with the proposal and development of automatic monitoring technology, automatic total station monitoring technology, optical fiber sensing technology, and 3D laser scanning technology have appeared. Digital close-up photography technology, monitoring technology based on the Internet of Things, etc., greatly improve the accuracy of monitoring and measurement, and provide reliable basic data for tunnel informatization construction.

Information-based design and construction technology is currently the mainstream of railway tunnel construction. Data on the geology, the surrounding rock dynamics, and the support status are obtained through advanced geological prediction, monitoring, and measurement. This approach makes it possible to sort and analyze the above-named data, assess the stability of the surrounding rock and support structure system, determine the support parameters, and construction methods that are more consistent with the surrounding rock dynamics, and guide on-site construction. In recent years, with the development of information-based design and construction technology, intelligent construction technology has also been proposed. In 2020, tests and verifications were successfully carried out around the major subsystems of the intelligent construction collaborative management platform in four sections of Zhengwan High-speed Railway, including Gaojiaping, Baokang, Xingshan, and Xinhua, and the purpose of intelligent tunnel construction was initially achieved [15, 16].

3.2 Tunnel auxiliary construction method of drilling and blasting method

The auxiliary construction methods commonly used in railway tunnels in China mainly include advanced bolting, advanced small conduit grouting, advanced pipe shed grouting, advanced grouting in the cave and surface grouting, etc. With the development of drilling and grouting equipment, the work efficiency of auxiliary construction methods has also been greatly improved, especially the development of advanced grouting in the cave and surface grouting technology, which provide important technical means for dealing with mud (water) intrusion in the tunnel, strengthening weak, broken strata and landslides and other unfavorable geology.

The advanced grouting technology in the cave has wide applicability, strong pertinence, and flexible operation. With the development of engineering requirements and technology, the current main processes include forward segmented grouting, drilling and injection integrated backward segmented grouting, horizontal sleeve valve tube bundle fine grouting, and full hole one-time constant pressure and constant flow grouting, which can be reasonably selected according to different geological characteristics and reinforcement requirements. Influenced by the effect and technology of advanced grouting in the cave, the deep hole grouting technology outside the cave has developed rapidly in recent years. This technology is widely used in the advanced reinforcement of various weak, water-rich, and broken rock tunnels. The construction depth of grouting holes can reach 150 m. After reinforcement, the average monthly excavation footage of soft rock tunnels can reach 80 m [17, 18]. The advanced grouting in the cave and surface grouting are shown in **Figures 4** and **5**.

The above technologies have been successfully applied in many projects such as Beijing-Shenzhen Passenger College, Ha-Mu high-speed railway, Menghua railway, Yinxi high-speed railway, Taijiao high-speed railway, Zhangjihuai railway, and so on.

In order to meet the needs of engineering grouting, the drilling and grouting machinery equipment have also made great development. The drilling equipment has been developed from the original tunnel drilling rig carried on the shoulder to the current crawler-type multi-functional drilling rig. The equipment is widely used in hydropower cofferdam construction, dam anti-seepage construction, highway slope anchoring, subway tunnel pipe shed support construction, airports, and deep foundation pits for high-rise buildings. Its construction efficiency and drilling capacity are

Figure 4.
Advance pre-grouting construction in the cave.

Figure 5.
Surface grouting construction.

greatly improved. At the same time, the grouting material has also developed from the traditional ordinary cement-based grouting material to the fast-hardening sulfate cement-based grouting material. This material has controllable setting time, high early strength, and strong dispersion resistance, which greatly improves the construction efficiency of advance grouting and the effect of water plugging and reinforcement.

In terms of grouting effect inspection and evaluation, on the basis of traditional analysis method, inspection hole method, excavation sampling method, displacement estimation method, etc., geophysical exploration methods such as in-hole imaging and cross-hole CT have been developed, so that the inspection of grouting effect is more intuitive and objective.

3.3 Mechanized large section construction technology of tunnel by drilling and blasting method

With the rapid development of high-speed railway in China, great progress has been made in the construction technology of high-speed railway tunnels. A new breakthrough has been made in the mechanized construction of drilling and blasting tunnels, and the mechanization of a single process has gradually shifted to the mechanized construction of all processes. The drilling and blasting method has evolved from manual blasting to drilling with rock drill jumbo or multi-arm drill truck. During the application of millisecond blasting, pre-crack blasting and smooth surface blasting technology, the entire excavation section is drilled, while blasting is either carried out simultaneously or at controlled and short intervals, to minimize vibration and other associated effects. The construction of mechanized large-section tunnels of Zheng-Wan high-speed railway and Cheng-Lan high-speed railway involved the construction of large-section excavations under surrounding rock conditions at all levels [7], as shown in **Figure 6**. The mechanization of the tunnel construction entails the main processes of operations such as excavation, initial support, waterproofing, and secondary lining. The main equipment includes rock drilling jumbo, bolt drilling rig, wet spraying manipulator, steel frame installation jumbo, self-propelled inverted arch trestle, etc., as shown in **Figure 7**. Rock drilling jumbo is a kind of rock drilling equipment used in tunnel and underground engineering. It can move and support multiple drilling machines to drill

(a)

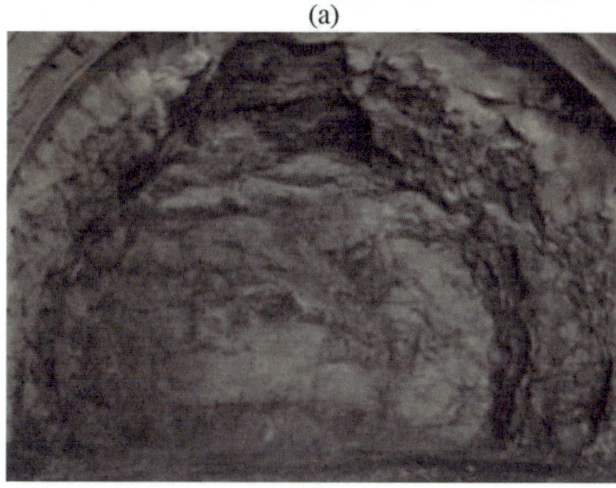

(b)

(c)

Figure 6.
Surface grouting construction. (a) Full-section method (including inverted arch). (b) Microstep method I (upper section inverted arch). (c) Microstep method II (including inverted arch).

simultaneously. Anchor Drilling Rigs roadway bolt support equipment is used to improve support effect, reduce cost, speed up roadway formation, reduce auxiliary transportation volume, reduce labor intensity, and improve roadway section utilization

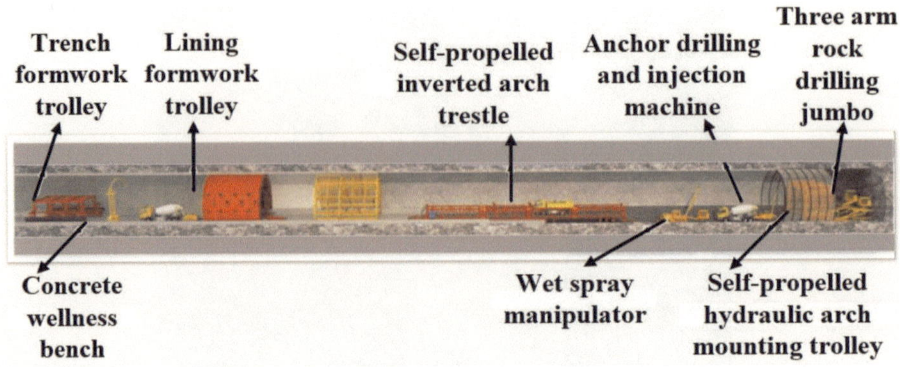

Figure 7.
Schematic diagram of the whole process of tunnel mechanical supporting.

rate. Wet spraying manipulator is a construction machinery designed and manufactured to replace manual spraying concrete in order to reduce labor intensity and improve working conditions. It has the functions of making the nozzle pitch forward and backward, swing left and right or circle, and the boom can be retractable, lifted and rotated to meet the construction requirements of shotcrete.

3.4 Tunnel construction technology of shield method

The shield method has strong adaptability to the stratum and can be applied to various strata such as clay, sand, bedrock, etc. It has obvious safety and economic advantages in the construction of urban railway tunnels and underwater railway tunnels. At the same time, in order to adapt to the complexity of geological conditions, shield equipment has been developed from the previous single-function earth pressure shield and mud-water shield to dual-mode and multi-mode shields.

The shield tunnel adopts the prefabricated segment assembly to form the lining structure in time, which ensures the safety, reliability, and high quality of the tunnel construction. The prefabricated assembly structure has the advantages of fast construction speed, good quality, and green environmental protection. Gradually popularize the application of prefabricated assembly structures, such as the Tsinghua Park shield tunnel of the Beijing-Zhangjiakou high-speed railway, and realize the fully prefabricated assembly of the under-track structure [19].

The segment is completed by the shield's own assembly equipment, and the bottom structure is completed by the matching box culvert assembly machine, realizing

Figure 8.
Box culvert and side culvert installation. (a) Central box culvert installation. (b) Side box culvert installation.

the rapid assembly construction of the middle-box culvert and the side box culvert. The middle box culvert block and the block and the side box culvert block and the block are all sealed with EPDM rubber strips. The installation of box culvert and side box is shown in **Figure 8**.

The tunnel constructed by this technology is clean and tidy in the tunnel, and the tunnel is completed at one time, which reduces the construction time of under rail structure and greatly improves the construction efficiency. It is the development direction of shield tunnel construction in the future [20].

3.5 TBM tunnel construction technology

The TBM method is a construction method for tunnel construction using a full-face tunnel boring machine. The hard rock TBM uses the disc cutter on the rotary cutterhead to squeeze and shear the rock, before picking up the stone slag using the bucket teeth on the rotary cutterhead. The slag is discharged to the main belt conveyor and transported it backwards, and then transports the slag to the outside of the tunnel through the traction slag truck or the tunnel continuous belt conveyor. It has the advantages of advanced technology, fast speed, one-time forming, good working environment, high safety, small disturbance to surrounding rock and small damage to natural environment. It enables the construction of deep buried long tunnel in complex geographical landform which is difficult to realize by traditional drilling and blasting method. It is one of the most advanced tunnel construction equipment [21]. China has successfully built Qinling Tunnel, Mogouling Tunnel, Taohuapu No. 1 Tunnel, Zhongtianshan Tunnel and West Qinling Tunnel by TBM technology. Through the joint efforts of domestic engineering construction, design, equipment manufacturing, and other related units, domestic TBM has been greatly improved in function, cost, reliability, automation, and geological adaptability, and can be applied to many fields of engineering and tunnels with different geological conditions. The full-face rock tunnel boring machine (TBM) has dominated the domestic and international market since the first domestic TBM was launched and successfully applied in 2014. It has laid a foundation for the wider application of TBM technology in railway mountain tunnel projects [22–24].

4. Development direction and prospect

The number and length of railway tunnels in China have ranked first in the world. The topography, geology, and regional environment are also the most complex in the world. The types, standards, and functions of tunnel engineering are complete, and the technical achievements are remarkable. However, with the high-quality development of China's economy and the application of new technologies such as informatization, digitization, and intelligence, the development of railway tunnel has also ushered in new opportunities and challenges.

4.1 Improving the active support system of tunnel

With the popularity of tunnel construction concepts such as "New Austrian Tunneling Method [2]," "Norwegian method [3]," and "New Italian method [4]," the development of railway tunnel design theory and method has been greatly promoted. The tunnel engineering community has gradually realized that protecting and making full use of the self-supporting capacity of tunnel surrounding rock is the core idea of tunnel construction. From the perspective of mechanics, the essence of tunnel support is to change the excavated surrounding rock from two-dimensional stress state to three-dimensional stress state, So as to inhibit the relaxation development of surrounding rock and improve the self-stability of surrounding rock.

Active support means that the support should be set up in time before the surrounding rock relaxes, and the surrounding rock should be actively protected, strengthened, and improved, mainly through anchor (cables) and other supporting components to penetrate into the surrounding rock to form a combined arch effect and improve the surrounding rock. It can improve the continuity of the surrounding rock and enhance the shear strength of the surrounding rock, thereby maintaining and improving the self-supporting ability of the surrounding rock. It is hoped that through the construction of national major railway tunnel projects and in combination with scientific research and tests, the concept of tunnel active support will be further improved and popularized, and the technical system of railway tunnel construction with Chinese characteristics will be improved.

4.2 Research on key technologies of BIM and GIS in tunnel life cycle

BIM (Building Information Modeling) takes the relevant information data of the construction project as the basis of the model, establishes the building model, and simulates the real information of the building through digital information simulation. It has the characteristics of information completeness, information relevance, information consistency, visualization, coordination, simulation, optimization, and plotting. At present, the three-dimensional simulation in BIM is mainly used to check the conflict in the construction process and improve the communication efficiency of project management [25]. With the popularization and application of BIM technology in the world, in recent years, China's survey and design units and construction units are also actively promoting the information construction based on BIM, and exploring the application of BIM Technology in the whole life cycle of Railway Tunnel Survey and design, construction, safety operation and maintenance. At present, it is leaping from the "modeling based" stage to the "multi-dimensional data application based" stage. However, BIM 3D information model has deficiencies in scale expression, consistent analysis, spatial unified benchmark, overall positioning and so on.

GIS (Geographic Information System) is based on spatial 3D visualization and spatial database technology. It is oriented to massive 3D geospatial data and integrates complete 3D spatial entities above, underground, and inside and outside the tunnel. It has powerful functions of spatial data storage, management, retrieval, and analysis. In addition, GIS can dynamically monitor and analyze environmental changes in different periods, and can also be developed to integrate data collection, spatial analysis, and decision-making processes into a common information flow. It significantly improves work efficiency and economic benefits, and provides technical support for solving urban tunnel construction problems and ensuring sustainable maintenance [26].

Therefore, combining BIM with GIS, integrating BIM model into GIS scene, and giving full play to their advantages can not only meet the accuracy requirements of different projects, but also reduce the measurement work, reduce the cost of spatial information collection, and realize the sharing and utilization of data [27]. This is the development direction of digitalization and informatization of tunnel engineering in the future [28].

4.3 Research on intelligent construction technology of railway tunnel

In recent years, with the continuous improvement of labor cost, the number of skilled construction technicians on the tunnel site has decreased year by year. The "machine instead of people" in tunnel engineering construction has become a realistic demand. Less people (or even unmanned) is the inevitable trend of tunnel engineering construction and development in the future. On the basis of the high integration of mechanization, dataization, informatization, and artificial intelligence, equipment (less internal combustion, more power, high efficiency, and capacity matching) that can adapt to the high altitude hypoxia environment is selected for hierarchical matching. At the same time, robots with self-perception, self-learning, self-decision, and self-implementation functions are developed to carry out intelligent operations in the main processes of tunnel construction. It is of great significance to build a full-life-cycle tunnel intelligent construction system with deep integration of mechanization and informatization to ensure safe and reliable construction [7, 16]. This is of great significance for improving construction efficiency, ensuring construction safety, and improving construction quality.

Based on the current demand for railway tunnel construction, technical level and development status, the development trend of railway tunnel intelligent construction technology is to implement intelligent construction system, continuously accumulate and improve tunnel design and construction methods under various geological conditions, and finally break through the theory of tunnel intelligent construction technology based on deep learning. Realize a self-learning and self-adaptive intelligent tunnel construction system. Subsequently, an intelligent tunnel construction system with dynamic perception, implementation analysis, accurate decision-making, and independent implementation is established to comprehensively promote and realize the intelligent tunnel construction.

At present, based on the mechanized large-section rapid construction technology of tunnel, the Hubei section of Zhengzhou Wanzhou high-speed railway has preliminarily constructed the intelligent construction technology system of high-speed railway tunnel, and has successfully carried out the site test.

4.4 New technologies for tunnel intelligent operation and maintenance

With the increase of railway tunnel engineering in China year by year, lining defects and diseases are also gradually increasing. The risk of safe operation of high-speed railway tunnel is prominent. The later tunnel operation and maintenance involve many problems, such as many personnel and equipment, complex road environment, high operation risk, difficult patrol inspection, etc. [29]. The maintenance demand of tunnel operation far exceeds the existing maintenance capacity. It is urgent to develop intelligent monitoring, operation and maintenance, and disease treatment technology of tunnel condition based on Internet of things technology [30]. It includes the research and development of structural health intelligent detection and monitoring system suitable for tunnel and underground engineering, tunnel condition evaluation intelligent decision system based on big data, efficient tunnel, and underground engineering maintenance technology and intelligent equipment.

The establishment of intelligent operation and maintenance system can improve the ability of information collection. With the help of mobile Internet technology and information standards, information transmission and integration can be accelerated. The use of artificial intelligence technology can enhance information mining capabilities and improve information collaboration capabilities. Thus, a set of information collection system, a set of network transmission system, a data processing center, an operation service platform, a collaborative service network, a set of technical and business standard system, and a set of responsibility traceability verification system can be formed. "Measurable, visible, controllable and usable" has been developed into "real-time monitoring, dynamic monitoring, intelligent control, timely service and accurate prediction" to provide guarantee for tunnel operation and maintenance management.

China has an extensive amount of tunnel construction in relation to most countries. The scale of tunneling is large. Based on years of research on operation and maintenance, the recommendation is to build a new "predictive" operation and maintenance system with full life cycle cost and performance as control indicators. At the same time, it is to develop integrated operation and maintenance technology equipment, rapid maintenance and reinforcement technologies, and multi-functional intelligent operation and maintenance management systems to achieve low-cost and highly efficient tunnel operation and maintenance management [30].

5. Epilogue

Over the past 40 years, the design method and construction technology of China's railway tunnel engineering have made rapid progress and made great achievements while drawing on the advanced technology at home and abroad. However, there are many technical problems when dealing with ultra-long and deep-buried tunnels, long-distance submarine tunnels, and those in complex urban environments. These problems are mainly focused on areas pertaining to intelligent construction, operation and maintenance. Therefore, railway workers should pay more attention to the intelligent construction of tunnels in future, to promote greater development of railway tunnel construction.

Author details

Siming Tian[1]*, Yang Xue[2], Zhenyan Tian[2], Shaoshuai Shi[1,2], Weidong Guo[2], Wei Wang[1] and Ruijie Zhao[3]

1 China Railway Economic and Planning Research Institute, Beijing, China

2 School of Qilu Transportation, Shandong University, Jinan, China

3 State Key Laboratory of Mining Disaster Prevention and Control Co-Founded by Shandong Province and The Ministry of Science and Technology, Shandong University of Science and Technology, Qingdao, China

*Address all correspondence to: tsmd@163.com

References

[1] Yong Z. Theory and Method of Tunnel Design. Beijing: China Communications Press Co., Ltd; 2019

[2] Wentian L, Jinhui C. Application of new Austrian tunnelling method to a large span trinocular cross section tunnel. Chinese Journal of Rock Mechanics and Engineering. 2017; **36**(11):2755-2766

[3] Zhilong W, Weihao W, Dagang L. Research on new tunnel support system based on Norwegian method Q system. Modern Tunnel Technology. 2022;**59**(3): 31-39

[4] Guohao Z, Cun Z, Jiajia G. Design and construction optimization of multi arch tunnel based on new idea method. Science Technology and Engineering. 2021;**21**(32):13914-13918

[5] State Railway Administration of the People's Republic of China. Code for Design of Railway Tunnel:TB10003–2016. Beijing: China Railway Press; 2017

[6] Mingnian W, Siguang Z, Xiao Z. Study of displacement control criterion for large-scale mechanized construction of tunnels on Zhengzhou-Wanzhou high-speed railway. Tunnel Construction. 2018;**8**:1271-1278

[7] Zhijian W. Innovation and future apllication of mechanized and intelligentized construction technology for high-speed railway tunnels: A case study of Hubei section on Zhengzhou-Wanzhou high-speed railway. Tunnel Construction. 2018;**3**: 339-348

[8] Siming T, Xu L, Dagang L. Theory and method of tunnel design under complex geological conditions. China Railway. 2020;**12**:43-50

[9] Mingqing X. Discussion on the design method of total safety factor of composite lining tunnel. Journal of Railway Engineering. 2018;**35**(1):84-88

[10] Mingqing X. Total Safety Factor Method for Tunnel Support Structure Design. Beijing: China communications Press Co., Ltd; 2020

[11] Mingqing X. Design of railway tunnel support structure based on total safety factor method. China Railway. 2020;**12**:82-88

[12] Shucai L, Bin L, Huaifeng S, et al. State of art and trends of advanced geological prediction in tunnel construction. Chinese Journal of Rock Mechanics and Engineering. 2014;**33**(6): 1090-1113

[13] Qihu Q. Main developments and directions of geological prediction and informatized technology of tunnel construction. Tunnel Construction. 2017; **37**(3):251-263

[14] Shucai LI. Theory and Method of Advance Geological Prediction of Water and Mud Inrush Disaster Sources in Tunnels. Beijing: Science Press; 2015

[15] Tongjun W. Development status and prospect of intelligent construction technology for railway tunnel in China. China Railway. 2020;**12**:1-9

[16] Feixiang L. Prospects for intelligent construction equipment technology innovation and collaborative construction management of railway tunnel. Tunnel Construction. 2019;**4**: 545-555

[17] Shijie L, Ze L, Jianhe W, Youjiang Z. Ground surface grouting techniques for the railway tunnel in completely

weathered granite stratum. Modern Tunnelling Technology. 2019;**6**:181-186

[18] Yonghu Z, Minglu B, Xinmin M, et al. Application research on the surface grouting in shallow buried loess tunnel with large section. Journal of Railway Engineering Society. 2019;**7**:48-51

[19] Yong Z, Gang L, Jianyou L, et al. Key technology innovation and application of Tsinghuayuan tunnel construction in Beijing-Zhangjiakou high-speed railway. Railway Standard Design. 2020;**64**(1): 109-136

[20] Youming Z, Zhiwei W, Zhihong W. Research and prospect on prefabricated structure of railway tunnel. China Railway. 2020;**12**:51-60

[21] Mengxue Q, Yanling Z. Research and application of SPS (continuous, balanced and fast) operation method in TBM tunnel construction. Tunnel Construction (Chinese and English). 2018;**38**(11):1860-1867

[22] Lijie D. Progresses, challenges and countermeasures for TBM construction technology in China. Tunnel Construction. 2017;**39**(7):1063-1075

[23] Kairong H, Dujuan W, Rujun G. Innovation and practice of hard rock TBM in China. Tunnel Construction. 2018;**38**(4):519-532

[24] Yanjun W, Mengxue Q. Prospects of key technologies of rock tunnel boring machine. Tunnel Construction. 2018; **38**(9):1428-1434

[25] Hao C, Yansong S, Huijie Z. Application of BIM+GIS in expressway intelligent tunnel monitoring platform. Highway. 2021;**66**(7):378-381

[26] Xiaorong Z, Xiaolu W, Can S. Application of UAV tilt photography technology combined with BIM+GIS in urban tunnel construction. Intelligent Building and Smart City. 2021;**12**:149-151

[27] Gaofeng H, Xianqi L, Hui Z, et al. Metro tunnel safety early warning and forecast technology based on the combined application of BIM and GIS. Urban Rail Transit Research. 2019;**22**(7): 161-164

[28] Wei N. Research on comprehensive application platform of design achievements of Xi'an-Shiyan High-speed railway based on GIS + BIM. Railway Standard Design. 2021;**65**(12): 1-5

[29] Yumei S, Yong L, Zhengang N. Application of 3DGIS and BIM integration technology in intelligent operation and maintenance of highway tunnel. Surveying and Mapping Bulletin. 2020;**10**:127-130

[30] Jianhong W, Quanyi X, Jian L, et al. Research on diseases and current situation of operation maintenance management of Japanese railway tunnels and suggestions. Tunnel Construction. 2020;**40**(12):1824-1833

Chapter 6

Underground Excavations below the Water Table by the Cut-and-Cover Method

Estanislao Pujades, Miao Jing, Chunhui Lu and Anna Jurado

Abstract

Most underground constructions, which are needed to improve mobility and increase available space in urban areas, require excavations that are usually deeper than the water table (e.g., for the construction of stations or underground parking lots). A frequently used technique to develop excavations under these conditions consists in combining the cut-and-cover method with a dewatering system based on deep pumping wells. Retaining walls used for the cut-and-cover method allow excavating between vertical walls and minimizes the inflow of groundwater, while deep pumping wells provide dry and stable conditions. Despite this technique is widely used, some aspects related with the presence of groundwater must be considered to avoid accidents. Dewatering systems must be properly designed to guarantee suitable conditions and to minimize the pumping settlements outside of the working area. In addition, it is required to assess the presence of defects in the retaining walls because the flow of groundwater through them may entail negative consequences. This chapter explains procedures (i) to design efficient dewatering systems considering the working conditions, the stability and the impacts generated in the vicinity of the construction, and (ii) to evaluate the state of the retaining walls by using hydrogeological tools.

Keywords: underground excavation, dewatering system, cut-and-cover method, pumping settlements, bottom stability, weather tightness assessment test

1. Introduction

The 70% of the world's population will be living in urban areas by 2050 [1]. Then, new infrastructures are needed for the development of cities and to cover the demands of the growing population. New infrastructures will improve the mobility (i.e., railway or motorway tunnels, stations, etc.) and the available space (i.e., underground parking lots). These new infrastructures must be constructed underground due to space limitations on the surface of urban areas, thereby, developing urban areas in the vertical direction. Most cities already have underground infrastructures and thus, the new ones should be deeper than the previous ones. Consequently, new infrastructures will probably be developed below the water table and will interact with groundwater.

The interaction between groundwater and the construction of underground infrastructures is bi-directional, meaning that the infrastructure may impact the natural behavior of aquifers (during and after its construction) and that the presence of groundwater hinders their construction, and in certain circumstances, even their utilization. Thus, the interaction between groundwater and underground infrastructures must be properly considered.

There are different construction methods adapted to the different kinds of infrastructures and the requirements of their construction. Among them, the cut-and-cover method is one of the most used in urban areas for constructing linear or non-linear infrastructures [2]. This method involves excavating between vertical retaining walls that minimize the needed space for the construction because, in the absence of retaining walls, the excavation walls must be inclined increasing enormously the needed space in case of deep excavations. The cut-and-cover method must be combined with dewatering facilities when excavations are developed below the water table [3]. Despite this dry conditions can be reached by collecting water at the surface by means of a system of trenches, and pumping it with a sump pump [4], stable conditions at the strata below the excavation bottom can only be reached by using deep pumping wells. The retaining walls (i) allow excavating between vertical walls, (ii) minimize the inflow inside the excavation and (iii) mitigate the potential impacts generated by the dewatering outside [5], whilst pumping wells reduce the piezometric head, ensuring dry conditions and providing stability to the excavation bottom [6]. The main actions to develop constructions by the cut-and-cover method combined with deep pumping wells (summarized in **Figure 1**) are: (1) to construct the retaining walls, (2) to drill the pumping wells and reduce the piezometric head until the required depth, (3) to excavate until the desired depth, (4) to construct the infrastructure and (5) to fill the space between the top of the infrastructure and the surface [7].

The presence of groundwater must be carefully considered during the execution of underground infrastructures by the cut-and-cover method. Groundwater may produce instabilities at the bottom of the excavation giving rise to liquefaction or bottom uplift events [8]. Then, the dewatering system, in addition to providing dry conditions, must reduce the water pressure as much as necessary to ensure stable conditions [6]. However, the hydraulic head should not be lowered more than necessary in order to minimize the impacts outside of the enclosure. The potential impacts produced

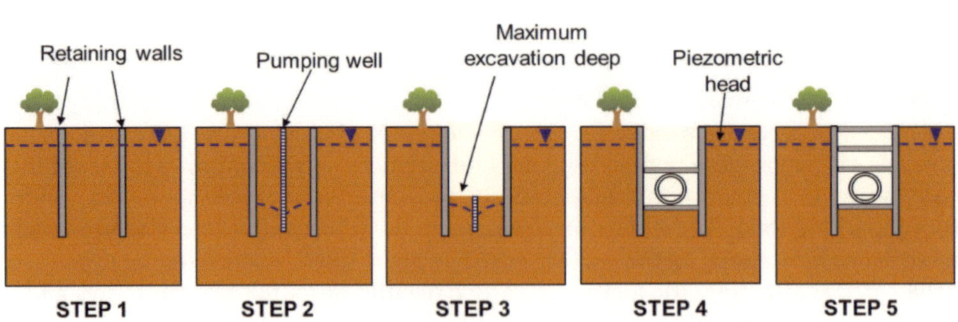

Figure 1.
Cut-and-cover method steps for the construction of an infrastructure below the water table (modified from [7]).

outside the enclosure are: (i) the loss of hydraulic resources [9], which is relevant in areas where water resources are scarce, and (ii) the soil subsidence (i.e., settlements) [10], which can endanger nearby buildings and infrastructures, either surface or underground. Common practices to minimize the groundwater-related issues consist of: (i) deepening the retaining walls until reaching an impervious layer and (ii) injecting jet-grouting below the excavation bottom to avoid the groundwater flow. However, both practices are risky since the stability of the excavation relies on the permeability of the retaining walls and the jet-grouting block. On the one hand, jet-grouting injection is complex and is difficult to reach a uniform and impervious block of treated soil [11, 12], especially in unconsolidated granular materials. On the other hand, defects in retaining walls, through which groundwater enters into the excavation, are relatively common [13, 14]. Then, the use of deep wells seems to be a more reliable option.

Faulty retaining walls do not block the groundwater and may endanger the construction development because they increase the risk of bottom instabilities, excavation flooding, sinkholes and soil deformations outside the enclosure [7, 15, 16]. Then, it is of paramount importance to determine the hydraulic conditions of the retaining walls before starting the excavation stage to adopt measures if needed (i.e., repair the retaining walls or redesign the dewatering system).

This chapter aims at explaining in detail how to design the dewatering system of deep excavations below the water table carried out by the cut-and-cover method considering groundwater stability issues and the potential impacts generated around the excavated area. In addition, this chapter exposes the consequences of faulty retaining walls and how they can be characterized by means of hydrogeological tools.

2. Dewatering system

The objective of a dewatering system is to provide workable conditions. Dry conditions are reached by dropping the piezometric head, at least, up to the maximum excavation depth, while stable conditions are achieved by reducing the water pressure in the soil located below the excavation bottom to guarantee that it is lower than the total vertical stress. Then, dewatering systems must be designed accounting for the piezometric head distribution above and below the maximum excavation bottom. The impact produced outside the excavation by the lowering of the piezometric head must be also considered when designing a dewatering system. The induced drawdown to ensure working conditions must not be excessive to minimize outer impacts such as (i) the loss of hydraulic resources, and (ii) soil settlements [17].

Groundwater can be extracted by different techniques but deep wells are the most appropriate in the context of urban deep excavations developed by the cut-and-cover method. Sump pumps [4] remove water from the excavation and allow working in dry conditions, but they do not drop enough of the water pressure below the excavation and bottom instabilities may occur. Well points are more adequate for shallow excavations because they extract water by suction and the maximum drawdown produced is lower than 6 m. Then, for developing deep excavations, well points must be installed in successive tiers or stages as excavation advances [18]. As a result, the area occupied by the excavation is usually very large, which is not acceptable in urban environments.

2.1 Bottom stability

Deep excavations developed below the water table can suffer bottom stability problems when pore water pressure (P_W) is too high. Bottom instabilities arise when P_W below the excavation bottom exceeds the total vertical stress (σ_V), then, according to Eq. (1) [19], when the effective vertical stress (σ'_V) is lower than 0,

$$\sigma'_V = \sigma_V - P_W, \tag{1}$$

where σ_V and P_W are defined as:

$$\sigma_V = z\gamma_S, \tag{2}$$

and

$$P_W = h\gamma_W \tag{3}$$

where γ_S and γ_W are the specific weights of the soil and water, respectively, z is the depth where σ_V is calculated, and h is the piezometric head at the same location. The two types of stability problems that excavation may suffer are liquefaction and bottom uplift (**Figure 2**) [8]. Liquefaction occurs when the excavation bottom is made of unconsolidated materials and under unconfined conditions. If σ'_V is lower than 0, the soil completely loses its shear strength and starts to behave like a fluid. Bottom uplift may occur when the excavation bottom is located above a confined aquifer whose P_W is not enough reduced [17]. Liquefaction and bottom uplift may also produce the deformation of the soil outside the excavation since the soil may migrate towards the centre of the excavation when they occur. In order to design dewatering systems to be able to guarantee the bottom stability, a safety factor (*SF*) is calculated below the excavation at different depths. Ideally, the depth of the points where *SF* is calculated should increase gradually. *SF* is defined as:

$$SF = \frac{\sigma_V}{P_W} \tag{4}$$

Stability is guaranteed when *SF* > 1. However, given that errors in the hydrogeological characterization are relatively common, it is advisable to consider

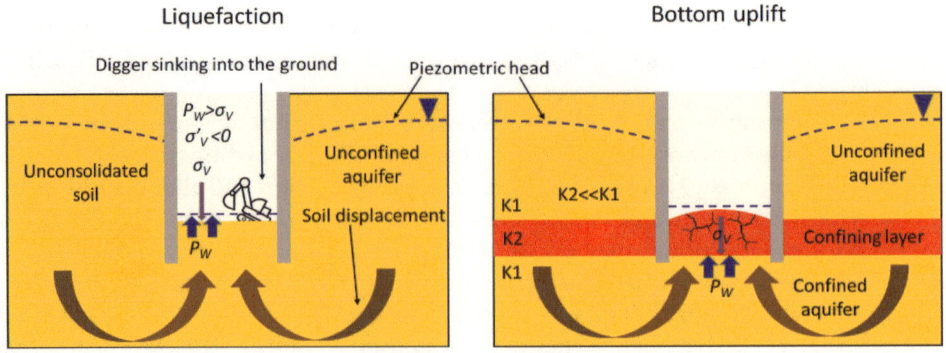

Figure 2.
Schematic description of bottom instabilities, liquefaction (left) and bottom uplift (right).

values of SF larger than 1 [20, 21]. Theoretically, a value of SF equal to or larger than 1.2 [22] should be sufficient as some strengths, like the friction between the soil and the retaining walls, that are opposite to P_W are not considered in Eq. (4). Consequently, SF computed using Eq. (4) is conservative. The stability of the bottom must be assessed for different dewatering schemes by numerically calculating the piezometric head and computing SF below the maximum excavation depth.

2.2 Impacts outside of the construction area

Dewatering systems may induce groundwater drawdown outside the excavation enclosure. Despite this impact is temporal, it must be considered in arid and semi-arid regions where water scarcity is a source of concern. Thus, dewatering systems must modify the piezometric head distribution outside of the enclosure as little as possible, which can be reached by deepening the retaining walls and avoiding oversizing the dewatering system more than needed to ensure stable conditions. Soil subsidence outside the enclosure as a result of the groundwater drawdown is another consequence of the dewatering. Soil subsidence (i.e., pumping settlement) is really feared in urban areas, the reason for which dewatering systems consisting of deep pumping wells are sometimes discarded. Instead, less efficient solutions are adopted like deepening the retaining walls more than needed until reaching deep low conductive layers [23] or using a jet-grouting bottom plug [24]. However, dewatering is not the only cause of soil subsidence occurring outside excavation; there are other construction-related actions that may induce larger soil deformations, like the digging of the retaining walls [20, 25]. Anyway, soil subsidence associated with the dewatering must be estimated during the design phase in order to choose the dewatering scheme that ensures more stability and less settlements. Pumping settlements can be calculated by using hydro-mechanical numerical models but they are time-consuming and a wide variety of parameters (hydraulic and mechanical) are needed, which are difficult to estimate. Therefore, analytical tools to predict the pumping settlements during the design phase in a short period and with acceptable error are of paramount importance.

Settlements can be computed by establishing the relation between the volumetric strain in the vertical direction (ε_z) and the vertical effective stress (σ'_z) as follows:

$$\varepsilon_z = \frac{\Delta z}{b} = \alpha \Delta \sigma'_V \tag{5}$$

where Δz is the length variation in the vertical direction (i.e., the settlement), b is the aquifer thickness and α is the compressibility of the porous material. Assuming a Biot's coefficient equal to 1, which is realistic for soft soils [26], then Eq. (1) is valid and σ_V can be considered constant. Thus,

$$\Delta z = \alpha b \Delta P_W \tag{6}$$

Eq. (6) is commonly used for the quantification of settlements caused by groundwater variations [17]. The meaning of α depends on the mechanical boundary conditions (i.e., displacement constraints). If the soil can only be deformed in the vertical (i.e., it is laterally confined), then α is the vertical compressibility of the soil [27] and is defined as [28–30].

$$\alpha = \frac{1}{(\lambda + 2G)} = \frac{(1+\nu)(1-2\nu)}{E(1-\nu)} \tag{7}$$

where G is the shear modulus, λ is the lame constant (drained conditions), E is the Young's modulus and ν is the Poisson's coefficient. If it is considered that displacements produced by groundwater drawdown occur in the three dimensions (e.g., in the presence of a pumping well [29, 30]), α can be defined as:

$$\alpha = \frac{1}{(\lambda + \frac{2}{3}G)} = \frac{3(1-2\nu)}{E} \tag{8}$$

Settlements around a single pumping well in steady state assuming a flat porous medium with axial symmetry can be computed from the solution proposed by Bear and Corapcioglu (1981) [30] as:

$$\Delta z = \frac{\rho_W g}{(\lambda + G)} \frac{Q}{4\pi K} ln \frac{R}{r} \tag{9}$$

where Q is the pumping rate, R is the influence radius of the pumping, r is the distance between the well and the site where the settlement is computed and K is the hydraulic conductivity of the porous medium ($K = T/b$, where T is the transmissivity of the aquifer. ρ_W and g are the water density and the gravitational constant, respectively, and are needed to express the variation of the groundwater head in Pascals. Note that Bear and Corapcioglu (1981) [30], defined α as $1/(\lambda + G)$ that is somewhat inconsistent with the common definition of compressibility shown in Eq. (8).

Previous equations assume that pumping settlements are proportional to groundwater drawdown, but this seems not to be true in the proximity of pumping wells as pointed out by Pujades [29] who modified Eq. (9) to improve the estimation of settlements occurring around pumping wells as follows:

$$\Delta z = \frac{\alpha \rho_W g Q}{4\pi K} = \begin{cases} ln\left(\frac{0.3b}{R}\right) & \text{for } r < \frac{9b}{70} \\ ln\left(\frac{0.3b}{R}\right) + \frac{ln^2\left(\frac{70r}{9b}\right)}{4\,ln\left(\frac{7}{3}\right)} & \text{for } r \geq 9b/70 \text{ and } r < 0.7b \\ ln\left(\frac{r}{R}\right) & \text{for } r \geq 0.7b \end{cases} \tag{10}$$

Assuming elastic behavior for aquifers, which occurs in the case of overconsolidation, the term $\alpha \rho_W g$ can be replaced by the specific storage coefficient (S_S) by considering the equation for elastic aquifers proposed by Jacob [31, 32]:

$$S_S = \rho_W g \alpha + \rho_W g \theta \beta \tag{11}$$

where θ is the porosity and β is the water compressibility coefficient. The first term of Eq. (12) ($\rho_W g \alpha$) is associated with soil deformation (S_{SE}) while the second term ($\rho_W g \theta \beta$) is related to water (S_{SW}). Considering that shallow-draining soils are much more compressible than water [33, 34], S_{SW} can be neglected. Thus,

$$S_S = S_{SE} = \rho_W g \alpha \tag{12}$$

Using S_S instead of $\rho_W g \alpha$ facilitates the computation of settlements since S_S can be easily derived from pumping tests. Parameters related to the compressibility of the soil can be also obtained from laboratory tests, but results may not be realistic since tested samples (i) cannot be representative of large volumes of aquifer and/or (ii) can be altered during the extraction processes.

Two kinds of settlements can be differentiated depending on their spatial distribution. If settlements are constant and their magnitude does not vary across the space, they are known as 'non-differential settlements', whilst 'differential settlements' occur when the soil deformation varies depending on the location. Potential damage of pumping settlements increases as induced settlements are more differential. Groundwater head distribution far from pumping wells is usually nearly flat, and therefore, large differential settlements are not expected from its variation. Largest hydraulic head variations are expected in the proximities of pumping wells, but, as observed by Pujades [29], pumping settlements close to pumping wells are constant and non-differential as a result of an arch effect that develops around the well.

However, this fact does not mean that pumping settlements are always non-differential and harmless because groundwater drawdown produces differential settlements in several situations, for example, when the aquifer is heterogeneous, which is relatively common. Variations in the hydraulic conductivity, the mechanical parameters or the geometry of the pumped aquifer induce differential settlements that can damage nearby infrastructures. This fact highlights the paramount importance of performing a detailed geological, hydrogeological and geotechnical characterization before designing dewatering systems.

3. Retaining walls

The objectives of the retaining walls are: (i) to allow excavating between vertical walls [5], (ii) to prevent the groundwater entrance into the excavation [35] and (iii) to minimize the impacts of the dewatering outside the enclosure [36]. There are different kinds of retaining walls that can be used for developing deep urban excavations below the water table such as retaining walls made up of concrete piles and panels or jet-grouting piles. Sheet piles can be also used for urban excavations but they are not used to support very deep excavations [37]. Retaining walls are made up of impervious materials to prevent the groundwater flow. However, construction defects in the retaining walls, through which groundwater can easily flow, are relatively common [13, 14]. There are several factors that may cause defects and they depend on the nature of the retaining walls. Defects in jet grouting retaining walls may be related to deviations and variations of the column dimensions [14], which is common in high vertical heterogeneity soils [38] because the area affected by the treatment depends on the properties of the soil. In addition, the presence of coarse sediments may produce shadow effects, leading to zones without treatment and openings [11]. The permeability of retaining walls made up of concrete piles or panels (i.e., diaphragm walls) can also be compromised by construction defects [21, 39] that can lead to openings. Openings may occur when: (i) sediments made of large boulders are drilled, (ii) the walls of the drilled space collapse or (3) the configuration of the slurry wall excavator is not suitable. Note that faulty enclosures commonly have numerous defects because, similar difficulties arise during the drilling of each one of the individual structures (i.e., concrete piles and panels or jet-grouting piles) that constitute an underground enclosure.

3.1 Consequences of faulty retaining walls

Consequences of faulty retaining walls depend on the relative location of the defects with respect to the bottom of the excavation (**Figure 3**). If defects are located above the excavation bottom, groundwater inflow (i) may drag sediments from outside leading to the formation of sinkholes [22] and (ii) may flood the excavation making it difficult for construction tasks. In addition, groundwater drawdown and its associated consequences (i.e., settlements) will increase outside the enclosure. If defects are located below the bottom of the excavation, drawdown and settlements will also increase outside the excavation site. Moreover, the pore water pressure will be higher than expected below the excavation leading to bottom instabilities (i.e., bottom uplift or liquefaction, structure instability and subsidence related to soil migration [15]).

3.2 Detection of defects

If retaining walls have defects, corrective measures can be carried out such as: (i) injecting a sealing substance to repair them, or alternatively, (ii) redesigning the dewatering system considering their actual hydraulic condition. In any case, the state of the retaining walls must be assessed before starting the excavation stage because corrective measures must be carried out before starting the pumping. If the excavation and the dewatering have started and there are attempts to repair the defects, the sealing substance will be dragged towards the pumping wells. As a result, the defects will not be repaired and the pumping wells may be damaged. If the dewatering system needs to be redesigned, new pumping wells should also be drilled before the excavation stage to minimize interference with the construction tasks and avoid bottom instabilities. Pujades et al. [7] and Pujades et al. [16] proposed two methods to assess the hydrogeological behavior of linear and non-linear underground enclosures and locate isolated defects. Both methods are based on pumping tests carried out inside the underground enclosure and their interpretation by using diagnostic plots. Both methods are based on the premise that changes in the flow behavior between linear and radial during the pumping depend on the characteristics of the retaining walls. Two different types of retaining walls are considered depending on the number and

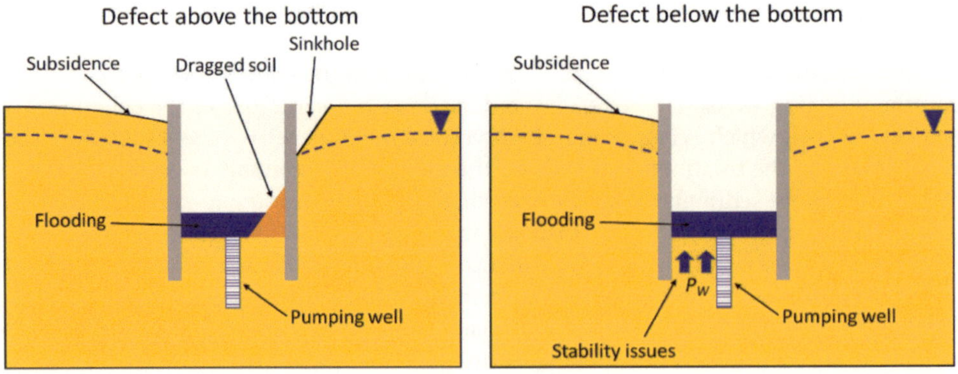

Figure 3.
Consequences of faulty retaining walls if defects are located above (left) and below (right) the bottom of the excavation.

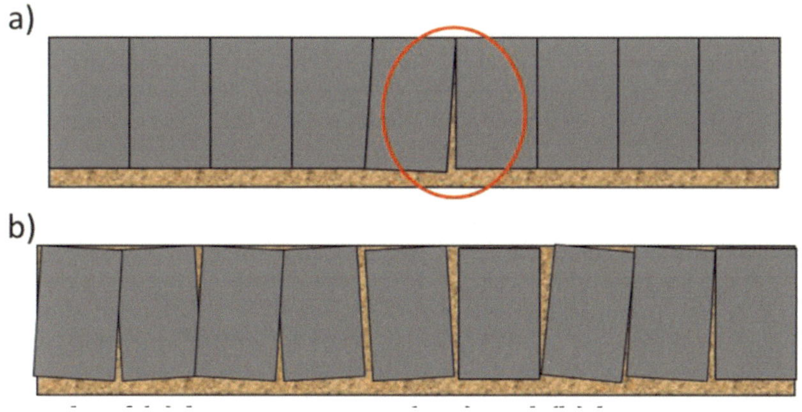

Figure 4.
Example of (a) heterogeneous and (b) homogeneous retaining walls.

distribution of defects (**Figure 4**). If there are isolated defects, retaining walls are named 'heterogeneous', while if there are numerous defects more or less homogeneously distributed, retaining walls are named 'homogeneous'.

3.2.1 Linear enclosures

Linear underground enclosures refer to underground excavations with a linear shape surrounded by retaining walls. This kind of enclosure is used, for example, for constructing tunnels. **Figure 5** illustrates the flow behavior evolution during pumping inside a linear underground enclosure. The flow is radial when the pumping starts because groundwater flows from all directions (**Figure 5a**). The flow is radial until the effect of the retaining walls reaches the pumping well, at this time, $0.1t_{L1}$, the behavior of the flow starts to change from radial to linear. The flow is totally linear at a time equal to t_{L1}. t_{L1} is defined as:

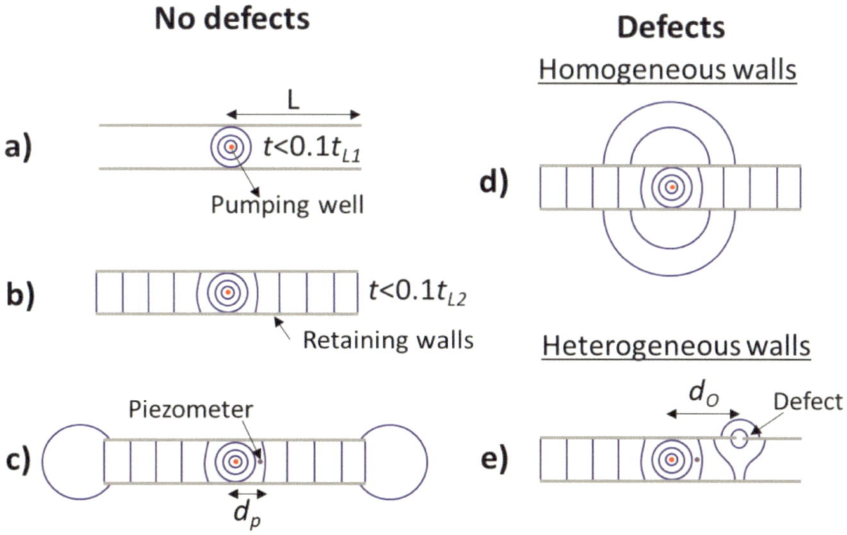

Figure 5.
Schematic description of the flow behavior during a pumping inside a linear enclosure.

$$t_{L1} = \frac{S(d_{dw})^2}{T} \tag{13}$$

where S is the storage coefficient of the aquifer, T is the transmissivity of the aquifer and d_{dw} is the distance between retaining walls. After t_{L1}, the flow is totally linear since pumped groundwater only arrives from the inner part of the enclosure that is drained linearly. If retaining walls are located at the edges of the enclosure, the flow continues to be linear until the enclosure is totally drained. However, if there are no retaining walls closing the edges of the enclosure, groundwater coming from all directions of the aquifer (i.e., with a radial behavior) enters the enclosure through them. Then, when the effect of the opened lateral edges of the enclosure reaches the pumping well the flow behavior at the pumping well changes from linear to radial. The behavior of the flow starts to change from linear to radial at a time equal to $0.1t_{L2}$ and it is totally radial after t_{L2}. t_{L2} is defined as:

$$t_{L2} = \frac{S(2L)^2}{T} \tag{14}$$

where L is the distance from the well until the end of the underground enclosure. If the behavior of the flow is observed in an observation point, t_{L2} is as follows:

$$t_{L2} = \frac{S(2L - d_p)^2}{T} \tag{15}$$

where d_p is the distance from the well until the observation point. The flow changes faster from linear to radial if there are defects in the retaining walls. If there are numerous defects and the retaining walls can be considered homogeneous (**Figure 4b**), the behavior of the flow starts to change from linear to radial at $0.1t_{L2.1}$, where $t_{L2.1}$ is defined as:

$$t_{L2.1} = \frac{S(1/\lambda)^2}{T} \tag{16}$$

and

$$\lambda = \sqrt{2\alpha_{dw}/Td_{dw}} \tag{17}$$

where $\alpha_{dw} = T_{dw}/w_{dw}$ is the leakage coefficient (conductance) of the retaining wall and T_{dw} and w_{dw} are the transmissivity and the thickness of the retaining walls, respectively. The inverse of λ is the length of retaining walls affected by the pumping when all pumped groundwater enters the enclosure through the defects [7]. Finally, if retaining walls have very few defects and can be considered heterogeneous (**Figure 4a**), the behavior flow changes from linear to a combination of radial and linear. The radial component is related to the groundwater entering the enclosure through the defect while the linear component is related to the groundwater coming from the side of the enclosure where the defect is not located. In this case, the flow does not behave as radial until a characteristic time equal to t_{L2}. The change in the flow behavior starts at a time of $0.1t_{L2.2}$, where $t_{L2.2}$ is:

DOI: http://dx.doi.org/10.5772/intechopen.109752

$$t_{L2.2} = \frac{S(2d_O)^2}{T} \tag{18}$$

d_O is the distance between the pumping well and the defect. If the head evolution is measured in an observation point different from the pumping well, $t_{L2.2}$ is:

$$t_{L2.2} = \frac{S\left(2d_O - d_p\right)^2}{T} \tag{19}$$

Identifying the changes in the flow behavior during a pumping test and applying Eqs. 13 to 19, it is possible to know (i) if the retaining walls are faulty or not, (ii) if they have numerous (homogeneous) or very few (heterogeneous) defects, (iii) their effective hydraulic conductivity and (iv) the position of the defects. However, flow behavior changes are really difficult to identify by observing the drawdown evolution. In this context, the use of diagnostic plots is useful to differentiate the changes in the behavior of the flow. Diagnostic plots are developed by calculating the derivatives of drawdown with respect to the logarithm of time (s') and plotting it versus time. s' is very sensitive to drawdown variations and, for this reason, diagnostic plots allow us to detect variations that are difficult to observe in the drawdown evolution [40].

Figure 6 shows the diagnostic plots calculated by considering infinite impervious retaining walls (black line), finite impervious retaining walls (purple line), faulty homogeneous retaining walls (red line), faulty heterogeneous retaining walls (blue dashed line) and the absence of retaining walls (gray line). The characteristics of the homogeneous and heterogeneous retaining walls have been chosen to ensure that their effective hydraulic conductivity is the same, and thus, 0.1 $t_{L2.1}$ and 0.1 $t_{L2.2}$ occur at the same time. In the beginning, the flow is radial for all scenarios since the retaining walls have not affected the pumping well. When it occurs, the flow behavior becomes linear and separates from the radial tendency (**Figure 6**, gray line). If the retaining walls have no defects (**Figure 6**, purple line), the flow is linear until the pumping well

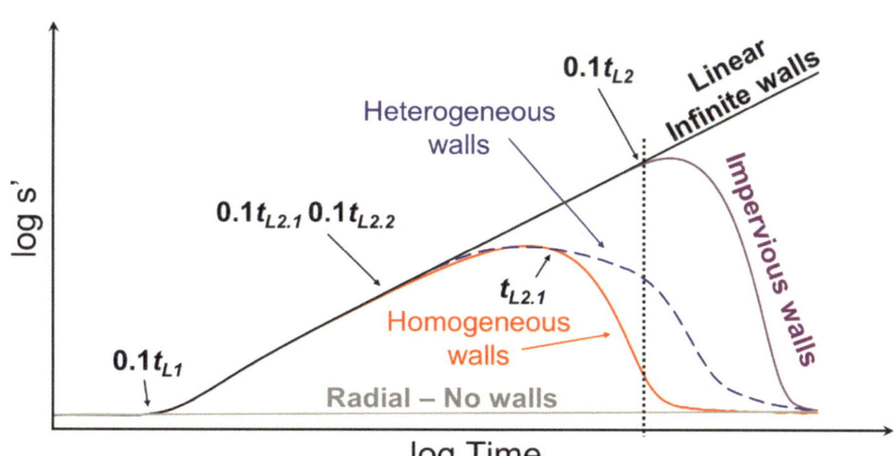

Figure 6.
Evolution of the logarithmic derivative of the drawdown (s') assuming infinite impervious retaining walls (black line), finite impervious retaining walls (purple line), faulty homogeneous retaining walls (red line), faulty heterogeneous retaining walls (blue dashed line) and the absence of retaining walls (gray line).

notices the effect of the end of the enclosure. At this time, *s'* separates from the linear tendency. If retaining walls have defects (**Figure 6**, red and blue dashed lines), the separation from the linear tendency occurs before. The transition period between linear and radial flow is shorter when retaining walls are homogeneous than when they are heterogeneous. If the retaining walls are homogeneous, the effect of radial flow starts to be observed at $0.1t_{L2.1}$ and s' is maximum at $0.5t_{L2.1}$, when half of the pumped flow is radial. This proportion increases with time and s' as the flow behavior becomes more and more radial. If the retaining walls are heterogeneous, the flow behavior separates from the linear tendency at $0.1t_{L2.2}$, but s' decrease more slowly than for homogeneous walls because the linear component of the continues being relatively high as groundwater from the aquifer (i.e., from all directions) only reaches the enclosure through the defect. The decrease rate of s' increases when the effect on the groundwater behavior of the lateral edges of the enclosure starts to be observed in the pumping well, then at $0.1t_{L2}$. **Figure 6** shows that *s'* has a different evolution depending on the kind of retaining walls. Once the retaining walls are classified as homogeneous or heterogeneous, it is possible to calculate their effective hydraulic conductivity (homogeneous retaining walls) or the position of the defect (heterogeneous retaining walls) only by identifying the times when s′ separates from the linear tendency.

Diagram in **Figure 7** aims at clarifying the changes in the flow behavior and the times when they occur. Difference in the flow behavior when the retaining walls are impervious or faulty occurs when the flow changes from linear to radial since t_{L1} is the same in all situations.

3.2.2 Non-linear enclosures

Non-linear underground enclosures refer to underground excavations with a non-linear shape surrounded by retaining walls. This kind of enclosure is used, for example, for constructing stations or emergency shafts for deep tunnels. **Figure 8a** illustrates the evolution of the groundwater flow behavior when pumping inside a non-linear enclosure and **Figure 8b** shows the evolution of the logarithmic derivative of the drawdown (*s'*) considering impervious retaining walls (black line), faulty homogeneous retaining walls (red line) and the absence of retaining walls (gray line). Groundwater flow behaves radially (stage 1) from the beginning of the pumping until the effect of the retaining walls is noticed at the observation point (e.g., the pumping

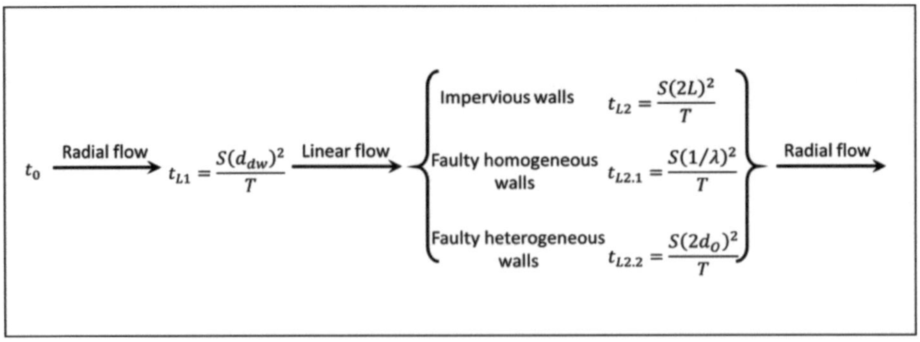

Figure 7.
Diagram of the flow behavior evolution for impervious and faulty retaining walls.

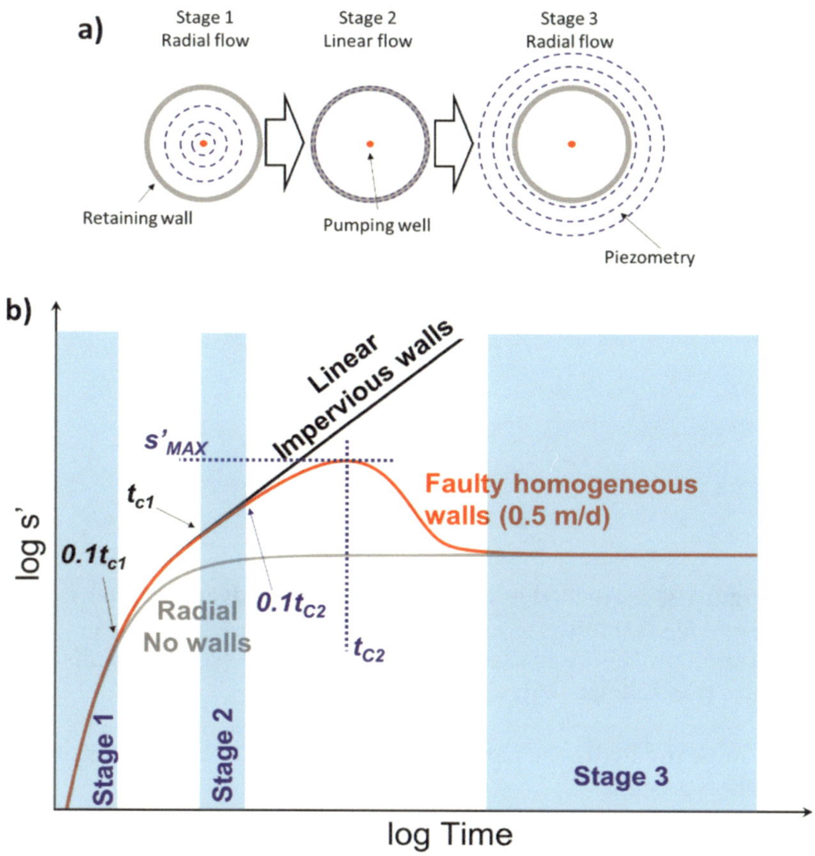

Figure 8.
a) Schematic description of the flow behavior during a pumping inside a circular (i.e., non-linear) enclosure. b) Evolution of the logarithmic derivative of the drawdown (s') assuming impervious retaining walls (black line), faulty homogeneous retaining walls with a hydraulic conductivity of 0.5 m/d (red line) and the absence of retaining walls (gray line).

well), when the flow behavior changes from radial to linear. This change in the flow behavior starts at $0.1t_{C1}$ and the flow is totally linear at t_{C1} defined as:

$$t_{C1} = \frac{S(2r_{exc})^2}{T} \tag{20}$$

where r_{exc} is the radius of the excavation. Note that t_{C1} may vary if the pumping well is not located in the centre of the excavation, but this effect is too short and can be neglected. In **Figure 8b**, $0.1t_{C1}$ occurs when the black and red lines separate from the gray line (i.e., the scenario without retaining walls where the groundwater flow is always radial). From t_{C1} to $0.1t_{C2}$, the flow is linear (stage 2) because the excavation is drained vertically and groundwater only comes from the inner part of the enclosure. The groundwater flow that crosses the retaining walls (Q_W) increases progressively as the piezometric head inside the enclosure decreases. Given that Q_W represents the portion of the groundwater crossing the retaining walls from the aquifer, it is a radial component controlling the flow behavior. Stage 2 finishes when the groundwater entering the enclosure through the retaining walls cannot be neglected and s' separates from the linear tendency (**Figure 8b**). After, there is a transition period until the flow behavior is totally radial (stage 3). During the first part of the transition period from stage 2 to stage 3, the pumped groundwater coming from inside the enclosure (Q_E) is higher than Q_W, and then, *s'* continues increasing. Note that Q_E is a linear component of the flow, and then, the flow behaves linearly as the portion of Q_E increases. During the second part of the transition period, Q_W is higher than Q_E and *s'* decreases. The inflection point in *s'* occurs when Q_W is equal to Q_E at t_{C2}. At this time, the maximum value of *s'* (s'_{MAX}) is reached. t_{C2} and s'_{MAX} are related to the effective hydraulic conductivity of the retaining walls (K_{eff}) as follows:

$$K_{eff} = \frac{S w_W r_{exc}}{2 t_{C2} b}, \tag{21}$$

and

$$K_{eff} = \frac{Q_p w_W}{2\pi r_{exc} s'_{MAX} b} \, ln\, 2.25 \tag{22}$$

Stage 3 (**Figure 8**) is reached when all pumped groundwater comes from outside the enclosure and Q_E is 0 m/d.

The application procedure to ascertain the state of the retaining walls in a non-linear enclosure is as follows [16]:

1. To determine the hydraulic parameters of the subsurface before starting the construction process.

2. To perform a watertightness assessment test (WTAT) [20, 41, 42] before starting the excavation and after drilling the retaining walls. The pumping well and piezometers must be located inside the underground enclosure, and at least, one observation point is needed.

3. To calculate *s'* from the observed drawdown and identify s'_{MAX} and t_{C2}. Depending on the location of the pumping and observation points, the number of observation points and the measured drawdown, different approaches can be adopted:

    a. K_{eff} can be always estimated from t_{C2} by using Eq. (21). Only one observation point is needed, but it is advisable to calculate K_{eff} using different observation points and compare the results.

    b. Pumping well and observation points are randomly distributed (**Figure 9a**): The pumping well is placed anywhere (i.e., it can be centred or not) and the piezometers are located at different distances from it. In this case, the retaining walls can be assessed from s'_{MAX} by using data from more than one observation point. If s'_{MAX} is equal in all observation points, the retaining walls behave homogenously and their K_{eff} can be computed by Eq. (22). If s'_{MAX} varies within observation points, (i) the enclosure is faulty and behaves heterogeneously or (ii) the retaining walls are homogeneous and their $K_{eff} \geq 0.01$ K. In this case, K_{eff} can only be estimated from t_{C2} by applying Eq. (21).

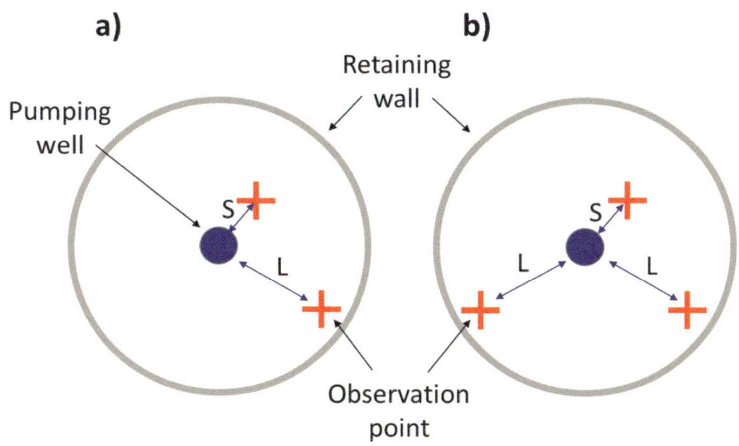

Figure 9.
Schematic description of the pumping well and observation points needed to characterize the state of a circular enclosure by a WTAT. ***Figure 9a*** *Shows a random distribution while* ***Figure 9b*** *shows a strategic distribution of all the required elements. S and L represent the distance from the well to the piezometers.*

c. Pumping well and observation points are strategically distributed (**Figure 9b**): The retaining walls can be assessed using s'_{MAX}. In this case, it is advisable to drill, at least, three piezometers and one centred pumping well. Two of the piezometers should be located equidistant from the pumping well whilst the other one should be drilled at a different distance.

   i. If s'_{MAX} is the same in all observation points, K_{eff} can be calculated by applying Eq. (22).

   ii. If s'_{MAX} is only equal in the observation points located at the same distance, the retaining walls are homogeneous and $K_{eff} \geq 0.01\, K$. Thus, K_{eff} cannot be estimated from s'_{MAX}.

   iii. If s'_{MAX} is not the same at any observation point, even at those located at the same distance from the pumping well, the retaining walls behave heterogeneously and K_{eff} cannot be estimated from s'_{MAX}. In the hypothetical and unlikely case that there is a defect located at the same distance to the observation points equidistant to the pumping well, s'_{MAX} will be equal at both observation points. Then, another piezometer at the same distance as the previous ones would be needed to establish the behavior of the retaining walls.

4. Conclusions

This chapter is focused on the interaction between deep excavations performed by the cut-and-cover method and groundwater. The first part of the chapter explains how to design dewatering systems to bring about workable conditions (i.e., dry and stable conditions) and estimate settlements generated outside the excavation as a result of lowering the piezometric head. Despite there being different kinds of dewatering systems, deep pumping wells seem to be the most adequate in the context of deep excavations. Deep pumping wells are useful to provide dry conditions, but also to reduce the water pressure below the excavation bottom bringing about stable conditions and avoiding liquefaction or bottom uplift events. The bottom stability must be anticipated during the design phase of dewatering systems by computing, for example, the safety factor as the ratio between the total vertical stress and the water pressure. The chosen dewatering system must be the one that, providing adequate conditions for the excavation, causes the minimum impacts outside of the construction area. Pumping settlements that occur as a result of lowering the piezometric head are really feared, especially in urban areas, because they can damage nearby infrastructures. Therefore, they must be estimated when designing dewatering systems. Hydromechanical models allow for estimating soil deformations precisely, but they are time-consuming and numerous parameters are needed. Thus, analytical equations like those proposed in this chapter are really useful to approximate the magnitude of the pumping settlements generated by a dewatering system.

The second part of the chapter describes how to ascertain the state of the retaining walls before starting the excavation phase. Retaining walls are essential to develop deep excavations in urban areas since they allow excavating within the enclosure of vertical walls (minimizing the needed space) and obstruct the flow of groundwater towards the excavation. However, defects in retaining walls are relatively common. If defects are not detected and remediation actions are not taken before starting the excavation phase, several issues may arise like the development of sinkholes in the outer part of the excavated enclosure, bottom instabilities or excavation flooding events. Here, two methods based on the interpretation of pumping tests are explained for the assessment of linear and non-linear enclosures. The methods allow to determine if the retaining walls have many defects or not. In the case where there are many defects, the hydraulic properties of the retaining walls can be known by identifying times when the behavior of the groundwater flow changes and applying the proposed equations. If there are few isolated defects, they can be located by analyzing the evolution of the groundwater flow behavior and using the proposed equations.

Acknowledgements

This research has been supported by the grant CEX2018-000794-S funded by MCIN/AEI/ 10.13039/501100011033, the grant RYC2020-029225-I funded by MCIN/AEI/10.13039/501100011033 and by 'ESF Investing in your future'. EP gratefully acknowledges the support received from the Barcelona City Council through the Award for Scientific Research into Urban Challenges in the City of Barcelona 2020 (20S08708) and the Spanish Ministry of Science, Innovation and Universities (MCIN/AEI /10.13039/501100011033/) and FEDER "one way to make Europe" through the project PID2021-128995OA-I00. M. Jing acknowledges the financial support from the National Natural Science Foundation of China (52109012).

Conflict of interest

The authors declare no conflict of interest.

Appendices and nomenclature

P_W	Water pressure
σ_V	total vertical stress
σ'_V	effective vertical stress
γ_S	specific weight of soil
γ_W	specific weight of water
z	soil/aquifer depth
h	piezometric head
SF	safety factor
ε_z	volumetric strain in the vertical direction
Δz	settlement – length variation in the vertical direction
b	aquifer thickness
α	compressibility of the porous material
G	shear modulus
λ	lame constant (drained conditions)
E	Young's modulus
ν	Poisson's coefficient
Q	pumping rate
R	influence radius of a pumping
r	radial distance from the pumping well
K	hydraulic conductivity
T	transmissivity of the aquifer
ρ_W	water density
g	gravitational constant
S_S	specific storage coefficient
θ	(porosity)
β	water compressibility coefficient
S_{SE}	portion of the specific storage coefficient associated with soil deformation
S_{SW}	portion of the specific storage coefficient associated with water
t_{L1}	time at which the behavior of the flow changes from radial to linear in the case of linear enclosures
d_{dw}	distance between retaining walls in liner enclosures
t_{L2}	time at which the behavior changes from linear to radial in the case of linear enclosures
L	distance from the well until the end of the linear underground enclosure
d_p	distance from the well until the observation point
$t_{L2.1}$	time at which the behavior of the flow changes from linear to radial behavior in case of faulty homogeneous retaining walls
$t_{L2.2}$	time at which the behavior of the flow changes from linear to radial behavior in case of faulty homogeneous retaining walls
α_{dw}	conductance of the retaining walls
T_{dw}	transmissivity of the retaining walls
w_{dw}	thickness of the retaining walls
d_O	distance between the pumping well and the defect
d_p	distance between the pumping well and the piezometer
s'	derivate of drawdown with respect to the logarithm of time
t_{C1}	time at which the behavior of the flow changes from radial to linear in the case of non-linear enclosures

r_{exc} radius of the excavation
Q_W pumped groundwater that reaches the inner part of the excavation by crossing the retaining walls
Q_E pumped groundwater coming from inside the enclosure
s'_{MAX} maximum value of s'
t_{C2} time when s'_{MAX} occurs

Author details

Estanislao Pujades[1*], Miao Jing[2,3], Chunhui Lu[3,4] and Anna Jurado[1]

1 Department of Geosciences, Institute of Environmental Assessment and Water Research (IDAEA), Severo Ochoa Excellence Center of the Spanish Council for Scientific Research (CSIC), Barcelona, Spain

2 School of Earth Science and Engineering, Hohai University, Nanjing, China

3 State Key Laboratory of Hydrology-Water Resources and Hydraulic Engineering, Hohai University, Nanjing, China

4 Yangtze Institute for Conservation and Development, Hohai University, Nanjing, China

*Address all correspondence to: estanislao.pujades@idaea.csic.es

References

[1] United Nations, Department of Economic and Social Affairs, Population Division. World Urbanization Prospects: The 2018 Revision. New York: UN; 2019

[2] Konda T. Shield tunneling method. Civil Engineering, Japan Society of Civil Engineers. 2001;**39**:23-27

[3] Forth RA. Groundwater and geotechnical aspects of deep excavations in Hong Kong. Engineering Geology. 2004;**72**(3):253-260

[4] Puller M. Deep Excavations: A Practical Manual. London: Thomas Telford; 1996. p. 474

[5] Xanthakos PP, Abramson LW, Bruce DA. Ground Control and Improvement. New York: John Wiley & Sons; 1994 938 p

[6] Linde-Arias E, Harris D, Davis A. Depressurisation for the excavation of Stepney green cavern. Proceedings of the Institution of Civil Engineers - Geotechnical Engineering. 2015;**168**(3): 215-226

[7] Pujades E, Carrera J, Vázquez-Suñé E, Jurado A, Vilarrasa V, Mascuñano-Salvador E. Hydraulic characterization of diaphragm walls for cut and cover tunnelling. Engineering Geology. 2012; **125**:1-10

[8] Jurado A, De Gaspari F, Vilarrasa V, Bolster D, Sánchez-Vila X, Fernàndez-Garcia D, et al. Probabilistic analysis of groundwater-related risks at subsurface excavation sites. Engineering Geology. 2012;**125**:35-44

[9] Guo F, Wang GH, Li ZC. Influence of artificial recharge in a phreatic aquifer on deep excavation dewatering: A case study of Dongguantou Nan Station in Beijing, China. Hydrogeology Journal. 2022;**30**(2):673-689

[10] Wu YX, Lyu HM, Han J, Shen SL. Dewatering—Induced building settlement around a deep excavation in soft deposit in Tianjin, China. Journal of Geotechnical and Geoenvironmental Engineering. 2019;**145**(5):05019003

[11] Vilarrasa V, Carrera J, Jurado A, Pujades E, Vázquez-Suné E. A methodology for characterizing the hydraulic effectiveness of an annular low-permeability barrier. Engineering Geology. 2011;**120**(1):68-80

[12] Pujades E, Vázquez-Suñé E, Carrera J, Vilarrasa V, De Simone S, Jurado A, et al. Deep enclosures versus pumping to reduce settlements during shaft excavations. Engineering Geology. 2014;**169**:100-111

[13] Bruce DA, De Paoli B, Mascardi C, Mongilardi E. Monitoring and quality control of a 100 metre deep diaphragm wall. In: Pilling and Deep Foundations. AA Balkema Rotterdam. 1989. p. 23-32

[14] Croce P, Modoni G. Design of jet-grouting cut-offs. Proceedings of the Institution of Civil Engineers - Ground Improvement. 2007;**11**(1):11-19

[15] Xu YS, Shen SL, Du YJ. Geological and hydrogeological environment in Shanghai with geohazards to construction and maintenance of infrastructures. Engineering Geology. 2009;**109**(3):241-254

[16] Pujades E, Jurado A, Carrera J, Vázquez-Suñé E, Dassargues A. Hydrogeological assessment of non-linear underground enclosures. Engineering Geology. 2016;**207**:91-102

[17] Cashman PM, Preene M. Groundwater Lowering in Construction: A Practical Guide. London: CRC Press; 2001. p. 498

[18] Shaqour FM, Hasan SE. Groundwater control for construction purposes: A case study from Kuwait. Environmental Geology. 2008;**53**(8): 1603-1612

[19] Terzaghi K, Peck RB, Mesri G. Soil Mechanics in Engineering Practice. New York: John Wiley & Sons; 1996. p. 579

[20] Pujades E, Vàzquez-Suñé E, Carrera J, Jurado A. Dewatering of a deep excavation undertaken in a layered soil. Engineering Geology. 2014;**178**: 15-27

[21] Wu YX, Shen SL, Xu YS, Yin ZY. Characteristics of groundwater seepage with cut-off wall in gravel aquifer. I: Field observations. Canadian Geotechnical Journal. 2015;**52**(10): 1526-1538

[22] Pujades E, Jurado A. Groundwater-related aspects during the development of deep excavations below the water table: A short review. Underground Space. 2021;**6**(1):35-45

[23] Baxter DY. Mechanical Behavior of Soil-Bentonite Cutoff Walls [Thesis]. USA: Virginia Tech; 2000

[24] Modoni G, Flora A, Lirer S, Ochmański M, Croce P. Design of jet Grouted Excavation Bottom Plugs. Journal of Geotechnical and Geoenvironmental Engineering. 2016; **142**(7):04016018

[25] Pujades E, Vázquez-Suñé E, Culí L, Carrera J, Ledesma A, Jurado A. Hydrogeological impact assessment by tunnelling at sites of high sensitivity. Engineering Geology. 2015;**193**:421-434

[26] Verruijt A. Consolidation of soils. In: Encyclopedia of Hydrological Sciences. Chichester, UK: John Wiley & Sons, Ltd; 2008

[27] Verruijt A. Theory of consolidation. In: Verruijt A, editor. An Introduction to Soil Dynamics. Dordrecht: Springer Netherlands; 2010. p. 65-90. DOI: 10.1007/978-90-481-3441-0_4

[28] Merxhani A. An introduction to linear poroelasticity. arXiv preprint. 2016; arXiv:160704274

[29] Pujades E, De Simone S, Carrera J, Vázquez-Suñé E, Jurado A. Settlements around pumping wells: Analysis of influential factors and a simple calculation procedure. Journal of Hydrology. 2017;**548**:225-236

[30] Bear J, Corapcioglu MY. Mathematical model for regional land subsidence due to pumping: 2. Integrated aquifer subsidence equations for vertical and horizontal displacements. Water Resources Research. 1981;**17**(4):947-958

[31] Jacob CE. Flow of Groundwater in Engineering Hydraulic. In: Rouse H, editor. Engineering Hydraulics. New York: John Wiley & Sons; 1950. pp. 321-386

[32] Ferris JG, Knowles DB, Brown RH, Stallman RH. Theory of aquifer tests. Theory of Aquifer Tests. Washington: U. S. Government Print Office; 1962, (Water Supply Paper; vols. 1536-E). Report No.: 1536-E. Available from: http://pubs.er.usgs.gov/publication/wsp1536E [Aceessed: April 4, 2022]

[33] Scott RF. Principles of Soil Mechanics. Reading, Mass: Addison-Wesley Pub. Co.; 1963

[34] Helm DC. Horizontal aquifer movement in a Theis-Thiem confined

system. Water Resources Research. 1994;**30**(4):953-964

[35] Wu YX, Shen SL, Yuan DJ. Characteristics of dewatering induced drawdown curve under blocking effect of retaining wall in aquifer. Journal of Hydrology. 2016;**539**:554-566

[36] Zeng CF, Song WW, Xue XL, Li MK, Bai N, Mei GX. Construction dewatering in a metro station incorporating buttress retaining wall to limit ground settlement: Insights from experimental modelling. Tunnelling and Underground Space Technology. 2021;**116**:104124

[37] Singh AP, Chatterjee K. Ground settlement and deflection response of cantilever sheet Pile Wall subjected to surcharge loading. Indian Geotech Journal. 2020;**50**(4):540-549

[38] Modoni G, Croce P, Mongiovì L. Theoretical modelling of jet grouting. Géotechnique. 2006;**56**(5):335-347

[39] Wu YX, Shen SL, Yin ZY, Xu YS. Characteristics of groundwater seepage with cut-off wall in gravel aquifer. II: Numerical analysis. Canadian Geotechnical Journal. 2015;**52**(10):1539-1549

[40] Renard P, Glenz D, Mejias M. Understanding diagnostic plots for well-test interpretation. Hydrogeology Journal. 2009;**17**(3):589-600

[41] Serrano-Juan A, Pujades E, Vázquez-Suñè E, Velasco V, Criollo R, Jurado A. Integration of groundwater by-pass facilities in the bottom slab design for large underground structures. Tunnelling and Underground Space Technology. 2018;**71**:231-243

[42] Serrano-Juan A, Pujades E, Vázquez-Suñè E, Crosetto M, Cuevas-González M, Leveling vs. InSAR in urban underground construction monitoring: Pros and cons. Case of la sagrera railway station (Barcelona, Spain). Engineering Geology. 2017;**218**:1-11

Chapter 7

On-Line Monitoring and Intelligent Diagnosis Technology of Rail Transit Ventilation System

Yongxing Song

Abstract

With the rapid development of economy and urbanization, subway has gradually become the main pillar of urban development. The ventilation system is the key guarantee of air quality in rail transit, and its condition monitoring and intelligent diagnosis are very important. The core problems of the complete set of a ventilation system required by the subway station have not been completely solved. The ventilation system includes the ventilator and additional equipment. The level of informatization and intelligence of the ventilation system and ventilator is not very high, and they have not yet been fully formed into an integrated diagnostic system. In view of the above two core issues, several scientific issues need to be tackled. This chapter studies the online monitoring and intelligent diagnosis mechanism of key equipment in the subway ventilation system. This mainly includes (1) modulation model of acoustic vibration signal; (2) noise reduction technology and feature extraction method; and (3) cases of multi-type typical fault identification fan equipment based on modulation model. Typical fault features were extracted respectively, which verified the effectiveness of the signal demodulation method for the diagnosis of rail transit ventilation systems.

Keywords: ventilation system, modulation model, feature extraction, fault identification, rail transit

1. Introduction

With the rapid development of economy and urbanization, rail transit has gradually become the main pillar of urban development. However, the underground tunnel is a relatively closed space. On the one hand, the ventilation system is needed to ensure the ventilation of the underground space to ensure the supply of fresh air for personnel and normal operation of equipment. On the other hand, in case of fire and other major fire-fighting failures, the normal operation of the fan is required to ensure the timely discharge of harmful gases and the inhalation of fresh air.

At present, the core problems of rail transit ventilation system have not been completely solved. The ventilator and ventilation system have not been fully formed as an integrated diagnostic system yet. The degree of informatization and intelligence of the ventilation system is not high. In view of the above core issues, several issues need to be tacked.

Multi-sensor data fusion technology is the first issue. There are hundreds of monitoring points in the ventilation system, which contain various types of sensor data. Multi-sensor data fusion is a multi-channel signal acquisition system that realizes the data fusion, data storage, and real-time display of multiple signals.

Then, fault diagnosis, prediction, and processing technology is the second issue. Based on the data integrated by the multi-sensor data fusion technology, the features will be extracted by signal post-processing methods. In a general case, the extracted features could represent fault feature, which are used for fault diagnosis and fault prediction.

Finally, the platform for data analysis and management is the third issue, which combines the multi-sensor data fusion technology with the fault diagnosis, prediction, and processing technology. The platform mainly involves sensor network technology, intelligent acquisition and monitoring technology, and network intelligent diagnosis technology. The fault diagnosis technology of rotating machinery, which has an impact on the precision of fault detection, is essential to the realization of the intelligent diagnosis technology of the rail transit ventilation system. Therefore, the fault feature extraction method of rotating machinery signals has been widely studied [1].

For the monitoring signals of rotating machinery, the traditional Fourier transform spectrum analysis method obtains frequency domain information, but time domain information is lost. The frequency domain analysis method can accurately obtain the characteristic information for the steady-state signal of rotating machinery, but in the non-stationary state condition, time-frequency analysis (TFA) is an effective method for signal feature extraction. Many researchers have conducted extensive research on TFA, mainly including short-time Fourier transform (STFT), wavelet transform (WT) [2], Wigner-Ville analysis [3] methods, and so on.

STFT is a TFA method developed based on Fourier transform, which can obtain the time-domain and frequency-domain information of the signal, and can realize better characterization of the non-stationary characteristics of the signal [4]. WT is an improvement of STFT, which overcomes the limitation of time and frequency resolution of STFT [5]. In addition, WT is also an effective signal noise reduction method. Liu et al. used WT to estimate the variance of vibration and noise signals, so as to realize the noise reduction of monitoring signals [6].

In order to improve the readability of TFA results, many signal decomposition and analysis methods have been proposed. Common signal decomposition methods include singular value decomposition [7, 8] (SVD), empirical mode decomposition [9, 10] (EMD), local mode decomposition [11] (LMD), and ensemble empirical mode decomposition [12] (EEMD). For the feature components of multi-component fault signals obtained by signal decomposition, principal component analysis (PCA) can be used as an effective data dimensionality reduction and feature extraction method [13–15]. Rahmani et al. [16] have proposed an efficient PCA algorithm. Li et al. [17] used the PCA method to extract multi-sensor features of nuclear power devices to realize fault detection, fault recognition, and feature reconstruction. Prawin and Rama Mohan Rao [18] used the PCA method to reconstruct the online time series input force signal and realized the extraction of principal component features. However, the feature extraction method based on signal TFA cannot characterize the modulation information of rotating machinery and thus cannot directly obtain the low-frequency modulation information.

During the operation of rotating machinery, mechanical resonance will be caused by the excitation force. The broadband mechanical vibration caused by excitation force is the carrier of the modulated signal. Then, the feature frequency of low-frequency modulation can be extracted by demodulating the resonant frequency band. Therefore, the demodulation algorithm based on resonance frequency band has been widely studied. Envelope demodulation (ED), Kurtogram, Fast Kurtogram (FK), and Protrugram algorithms have been proposed successively, which are widely used in the extraction of fault features of rotating machinery or components [19]. Under the situations where the signal-to-noise ratio (SNR) is good, the spectral kurtosis (SK) analysis algorithm based on resonance narrowband signal demodulation can obtain better demodulation results. However, under the interference of strong non-Gaussian background noise, the SK analysis demodulation algorithm is easy to fail. When the SNR of the monitoring signal decreases, the demodulation and analysis performance based on the resonant frequency band decreases rapidly.

The demodulation method based on high-order statistics is an effective method to extract fault features of rotating machinery with high accuracy. It has better demodulation accuracy than the narrowband demodulation algorithm based on resonance band demodulation. Cyclostationary demodulation algorithm is a typical algorithm based on high-order statistics demodulation. Antoni and Randall [20] analyzed the relationship between correlation spectrum and envelope spectrum, and the research results show that correlation spectrum has better fault feature extraction accuracy. The analysis method based on second-order cyclostationarity is the most effective method to extract the fault information of rotating machinery from high-order statistics. Antoni [21] have carried out a detailed application research on the analysis method of cyclostationary signals, which have a good feature extraction effect on traditional rotating machinery.

In the research of rotating machinery demodulation algorithm, it is generally assumed that the noise interference signal is Gaussian white noise. Nevertheless, it is not consistent with many practical application scenarios. Borghesani et al. [22] carried out demodulation analysis and research on cyclostationary signals under non-white noise interference and proposed square envelope spectrum. However, the spectral correlation (SC) analysis algorithm has high computational complexity. Therefore, the SC analysis method cannot be effectively used to realize the online monitoring and fault diagnosis of rotating machinery. Antoni et al. [23] proposed a Fast-SC analysis method based on STFT, which was verified and analyzed by bearing fault experiments. Horstmann et al. [24] proposed the detection and identification of approximate cyclostationary signals and the estimation method of a cycle period. In the algorithm, signal resampling is the key technology. Sophie et al. [25] proposed to use the cyclostationary analysis method in the angle/time domain to identify the fault characteristics of the bearing housing under unsteady working conditions. The key is to use the encoder to collect the phase information and resample the time-domain fault vibration signal monitored by the rotating machinery. Borghesani and Antoni [26] analyzed the failure performance of square envelope spectrum and cyclic demodulation spectrum under peak background noise and further analyzed the effectiveness and anti-noise of logarithmic envelope spectrum by simulation analysis and experimental verification.

According to the research status of demodulation algorithm based on high-order statistics, although the demodulation algorithm based on high-order statistics can obtain better demodulation accuracy, its computational efficiency is low, and its noise resistance needs to be further improved, so it is difficult to realize the online analysis of monitoring signals. For these defaults, the DPCA method, which is a demodulation method based on TFA and PCA, was proposed by Song et al. [27, 28]. Based on this method, the dimension of time-frequency distribution matrix can be reduced to realize the fast demodulation of signals. The computational efficiency of this algorithm is sufficient to realize the online analysis of monitoring signals.

2. Modulation models of acoustic vibration signal

Considering the working characteristics of rotating machinery, during the operation process, due to rotor imbalance, misalignment, flow field instability, and other factors, periodic impact will occur, resulting in significant modulation signal components in the radiated noise of fan equipment. Therefore, according to different working conditions of rotating machinery, its modulation model can be divided into amplitude modulation (AM) signal model under steady-state conditions and amplitude modulation-frequency modulation (AM-FM) signal model under unsteady state conditions. The two working conditions are respectively for constant speed operation and variable speed operation. In this section, typical modulation models of above modulation signals will be illustrated and analyzed, respectively.

2.1 Modulation model of AM signal

Under the condition of constant speed operation, the radiated noise signal produced by rotating machinery contains obvious amplitude modulation signal, which is mainly due to its periodic excitation force. Under steady-state conditions, because the running speed of the rotating machinery remains unchanged, the action cycle of impact force remains unchanged, and finally the characteristic modulation frequency of amplitude modulation signal remains stable.

$$x_{MA}(\mathrm{t}) = A_m \cos\left(2\pi f_m \mathrm{t}\right) \sum_{i}^{N} \cos\left(2\pi f_{c,i} \mathrm{t}\right) \tag{1}$$

where x_{MA} is the AM signal of rotating machinery; A_m is the amplitude of the AM signal; f_m is the characteristic frequency of the AM signal; $f_{c,i}$ is the frequency of the carrier signal; N is the total number of carrier signals.

When the spectrum of the carrier signal is line spectrum, its envelope signal has significant periodicity, and its carrier modulated AM signal has obvious spectral characteristics, and the sideband has obvious symmetry, as shown in **Figure 1**. Such features can accurately reflect the characteristic frequency information of the modulated signal. Therefore, for a simple single component AM modulation signal, envelope demodulation, resonance demodulation, spectral kurtosis analysis, and other algorithms can be used to extract the characteristic frequency information of the modulation signal in the monitoring signal.

When the carrier signal spectrum is broadband signal, the envelope signal and spectrum of broadband carrier AM simulation signal are shown in **Figure 2**. The time-domain and frequency domain spectrum characteristics of the amplitude modulation signal are different from those of the single component line spectrum carrier modulation signal.

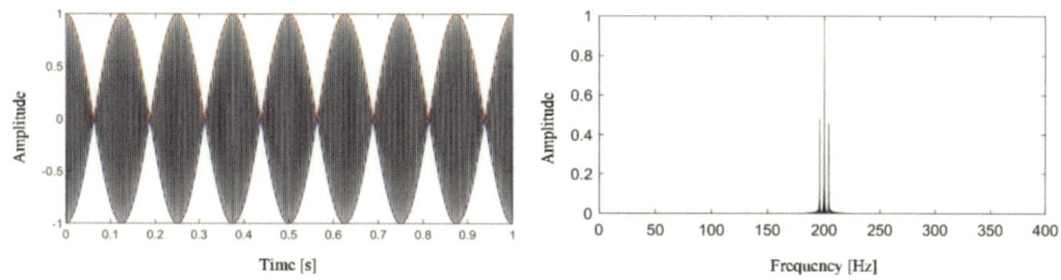

Figure 1.
Mono-component AM signal and its spectrum [29].

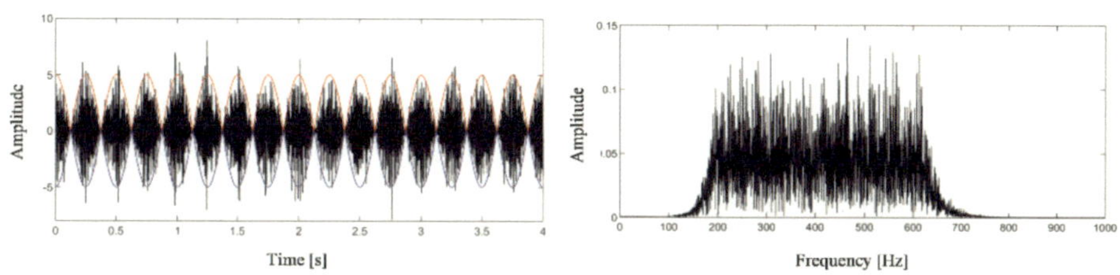

Figure 2.
Wideband carrier AM signal and its spectrum [29].

In time-domain, the envelope signal of wideband carrier modulation signal has deviation, but the overall trend corresponds to the modulation signal, so it can reflect the characteristic modulation information of amplitude modulation signal. In the frequency domain, the spectrum characteristics of wideband carrier modulation simulation signal appear as wideband frequency domain signal. At this time, there is no significant distribution feature in the spectrum of the amplitude modulated signal, so the signal can be demodulated by using the envelope spectrum analysis method, but the characteristic frequency information of the modulated signal cannot be observed directly from its spectrum.

2.2 Modulation model of the AM-FM signal

Under the condition of variable speed operation, the amplitude and frequency of the radiated noise signal generated by rotating machinery are modulated at the same time. At this time, the modulation signal is mainly amplitude modulation frequency modulation signal. The main reason for this signal is the periodic change of the impact force of the rotating machinery with time in the process of variable speed operation. The AM-FM signal of the radiated noise signal of the rotating machinery established in this paper is as follows:

$$x_{MAf}(\mathrm{t}) = A_{mf}\cos\left(\theta_{mf}(\mathrm{t})\right)\sum_{i}^{n}\cos\left(2\pi f_{i}\mathrm{t}\right) \tag{2}$$

where x_{MAf} (t) is the AM-FM signal; A_{mf} is the amplitude of the AM-FM signal; θ_{mf} (t) is the change of the modulated signal angle; f_{mf} (t) is the instantaneous frequency of the AM-FM signal, as expressed in Eq. (3).

$$f_{mf}(\mathrm{t}) = \frac{d\theta_{mf}(\mathrm{t})}{dt} \tag{3}$$

For a typical AM-FM signal, the envelope signal of its timing signal is different from that of a typical AM signal, and its envelope signal does not have typical periodicity. When the carrier signal is a line spectrum carrier signal, the timing waveform of the single component AM-FM signal and its corresponding spectrum distribution are shown in **Figure 3**.

According to the waveform of single component AM-FM signal, at this time, the envelope signal can still reflect the waveform characteristics of the modulation signal, but because the frequency of the modulation signal changes, its waveform does not have periodicity. Different from the spectrum of single component am line spectrum carrier modulation signal, the spectrum of single component AM-FM signal does not have obvious sideband effect and presents a certain bandwidth as a whole, so the corresponding modulation information cannot be obtained according to its spectral characteristics.

When the carrier signal is broadband noise, the time-domain waveform and spectrum distribution of the AM-FM simulation signal of the broadband carrier signal are shown in **Figure 4**. The waveform characteristics of its envelope signal are similar to the envelope signal of single component AM-FM line spectrum carrier modulation signal, which can reflect the waveform of the modulation signal, but there is a certain error. The spectrum of the wideband carrier modulated signal as a whole is a wideband frequency domain signal, which corresponds to the wideband spectrum of the carrier signal as a whole, but it does not reflect the change of the characteristic frequency of the modulated signal. Therefore, it is difficult to extract the characteristics of AM-FM signals using conventional envelope demodulation algorithm.

For the feature extraction of AM-FM signal, the signal resampling technology and demodulation technology are often used. The resampling operation of the monitoring signal of rotating machinery requires the monitoring of phase information. The resampling signal of rotating machinery can be obtained by using the phase information of rotating machinery for data difference and combining the timing signal of the

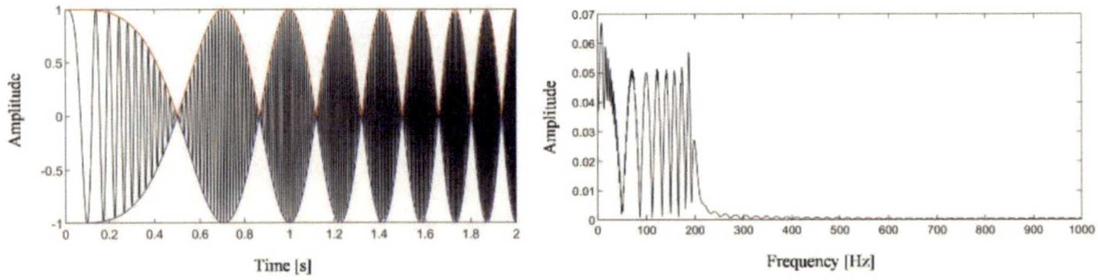

Figure 3.
Mono-component AM-FM signal and its spectrum [29].

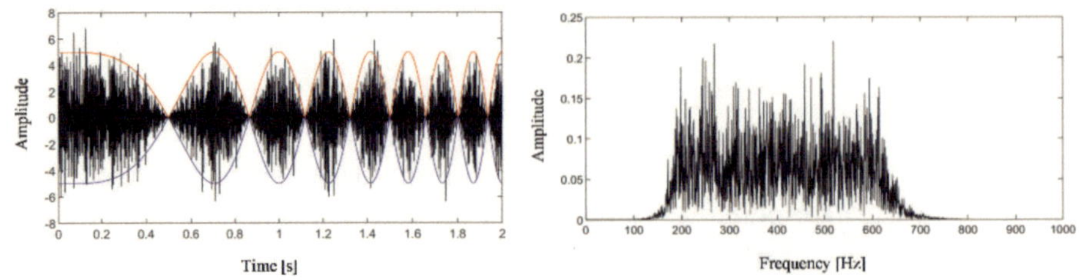

Figure 4.
Wideband carrier AM-FM signal and its spectrum [29].

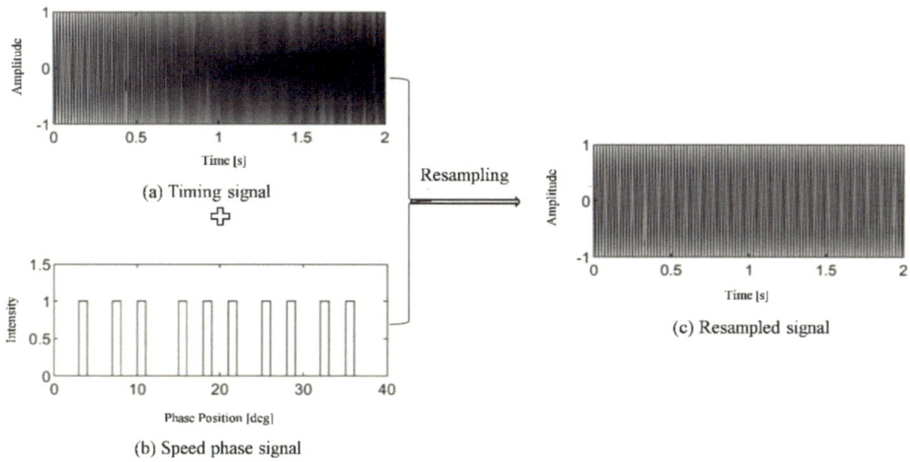

Figure 5.
Monitoring signal resampling [29]. (a) Timing signal; (b) speed phase signal; and (c) resampled signal.

monitoring signal, as shown in **Figure 5**. According to the resampled signal obtained by resampling the phase data of rotating machinery and the radiated noise signal, combined with the method of spectrum analysis, the order modulation information of the acoustic signal of rotating machinery can be obtained.

3. Noise reduction technology and feature extraction method of online fan monitoring signal

3.1 Radiated noise signal components

The acoustic signal of rotating equipment mainly includes three components: deterministic signal, modulated signal, and noise signal [29]. Therefore, the sound radiation model of rotating machinery can be expressed as **Figure 6**. The vibration transmission system of rotating machinery can be regarded as a liner time invariant system. The monitoring signal is the convolution of above three signal components and transmission path function, as illustrated in Eq. (4)

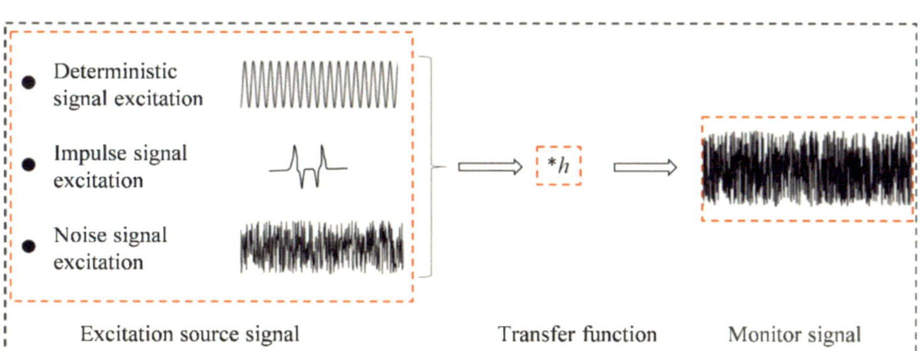

Figure 6.
Acoustic signal model of rotating machinery [29].

$$x(t) = [x_d(t) + x_m(t) + x_e(t)] * h \tag{4}$$

where $x(t)$ is monitoring signal; $x_d(t)$ is deterministic signal component; $x_m(t)$ is modulation signal component; $x_e(t)$ is noise signal component; * represents convolution operation; h is the transfer function from the vibration source to the monitoring point. It is worth noting that the impact of transfer function can be ignored since the transfer function only has a small influence on the frequency characteristics of signal.

The deterministic signal components in the monitoring signal are caused by mechanical faults in the process of mechanical operation, such as rotor imbalance, bolt looseness, and so on. The deterministic signal components can be accurately characterized by functional models, which have first-order statistical characteristics under steady-state conditions as shown in Eq. (5).

$$x_d(t) = E[x_d(t + T_d)] \tag{5}$$

where, E [] represents the statistical average function of the signal; T_{d} represents the period of the signal.

Under steady-state condition, the deterministic signal component is the basis of signal preprocessing analysis since the characteristic frequency and amplitude of the component do not change considerably.

Modulation signal component is the characteristic component in the monitoring signal. The modulation signal component is caused by the periodic impulse of rotating parts in the process of rotation. Under constant speed operation, the modulation signal is AM signal, whose second-order statistical characteristics have significant periodicity as shown in Eq. (6).

$$x_{cr}(t, \tau) = E[x_m(t)[x_m(t + \tau)]] \tag{6}$$

where $x_{cr}(t, \tau)$ is the autocorrelation function of the rotating machinery monitoring signal; τ is the delay time.

Under variable speed operation, the modulation signal is AM-FM signal; the relationship between the order of monitoring signal and the frequency modulation can also be obtained by resampling. In rotating equipment, the fault features are hidden in the modulation signal. Faults of various positions, components, and degrees exhibit distinct characteristics in modulated signals. There is a correlation between fault features and modulation features. Thus, the fault early warning and fault location could be realized. Therefore, the modulation signal components in radiation noise generated by rotating machinery have periodicity of high-order statistics. It is an effective means for the application of rotating machinery condition monitoring, fault early warning, and fault location.

Rotating machinery noise signal components are mainly Gaussian white noise, which does not have the periodicity of first-order, second-order, and higher-order statistics. Therefore, in order to accurately extract the low-frequency acoustic texture feature information of rotating machinery, it is necessary to eliminate noise signal component by utilizing its statistical characteristics.

3.2 Signal preprocessing method

Under the condition of low SNR, the interference of environmental noise and equipment noise leads to serious clutter interference in the radiated noise signal, which is not conducive to the identification and extraction of deterministic characteristic frequencies. The radiation noise is composed of equipment noise and environmental noise. The equipment noise is the noise generated by equipment operation, which refers to the noise with a certain modulated signal component. The environmental noise is the background noise, which refers to noise unrelated to the operation of equipment. The equipment noise contains deterministic signal component and the modulation signal component, while the environmental noise contains noise signal. Signal preprocessing method is an effective noise reduction method for monitoring signals. The preprocessing methods for rotating machinery monitoring signals mainly include time-domain synchronous averaging (TSA) method and auto-regression (AR) model.

3.2.1 Time-domain synchronous averaging

Because the modulated signal components and noise signal components do not have first-order steady-state statistical characteristics, while the deterministic signal components have stable first-order statistical characteristics, synchronous averaging is a commonly used deterministic component extraction and noise reduction method for rotating machinery equipment running at a constant speed, as shown in Eq. (7). TSA is the most commonly used synchronous averaging technology.

$$x_{TSA}(\mathrm{t}) = \lim_{M\to\infty} \frac{1}{2M+1} \sum_{m=-M}^{M} x(\mathrm{t} + mT_{TSA}) \tag{7}$$

where $x(\mathrm{t})$ is the monitoring signal; $2M+1$ is the number of averaging; T_{TSA} is the duration of averaging period; $x_{\mathrm{TSA}}(\mathrm{t})$ is the TSA signal.

However, under the variable speed operation, the TSA cannot be directly used. The monitoring signal needs to be resampled by the phase signal monitored by the rotating machinery. According to the resampled signal, combined with the angle-domain average processing, the angle-domain averaging signal can be obtained，as shown in Eq. (8).

$$x_{TSA}(\theta) = \lim_{M\to\infty} \frac{1}{2M+1} \sum_{m=-M}^{M} x(\theta + m\theta_{TSA}) \tag{8}$$

where $x(\theta)$ is the resampling monitoring signal; θ_{TSA} is the angle averaging period; $x_{\mathrm{TSA}}(\theta)$ is the angle-domain averaging signal.

However, there are a series of problems in the practical application of TSA. In practice, the feature extraction of deterministic signal components needs to have certain requirements for the resolution of time-domain signals. Moreover, in order to meet the requirements of signal analysis frequency resolution, TSA has higher requirements for the temporal length of monitoring signals and the operating conditions of rotating machinery, which play a certain role in limiting the application of the TSA technology.

3.2.2 Linear prediction based on auto-regression model

Linear prediction is an effective algorithm for extracting the deterministic components in signals. This algorithm can use historical data to achieve accurate prediction of data and then realize the prediction and separation of deterministic components in rotating machinery monitoring signals. The AR model used to extract the deterministic components can be expressed by Eq. (9).

$$x_{AR} = -\sum_{i=1}^{p} q(i)x(n-i) \tag{9}$$

where x_{AR} is the deterministic signal component obtained by linear prediction of monitoring signal; p is the order of AR model; $q(i)$ is weight coefficient. $q(i)$ can be solved by Yule-Walker equations. The process is as follows:

$$\begin{bmatrix} r_{xx}(0) & r_{xx}(-1) & \cdots & r_{xx}(-p+1) \\ r_{xx}(1) & r_{xx}(0) & \cdots & r_{xx}(-p-2) \\ \vdots & \vdots & \ddots & \vdots \\ r_{xx}(p-1) & r_{xx}(p-2) & \cdots & r_{xx}(0) \end{bmatrix} \begin{bmatrix} q(1) \\ q(2) \\ \vdots \\ q(p) \end{bmatrix} = - \begin{bmatrix} r_{xx}(1) \\ r_{xx}(2) \\ \vdots \\ r_{xx}(p) \end{bmatrix} \tag{10}$$

The residual signal refers to an interference signal, which is composed of modulation signal and noise signal. The residual signal was obtained by the residual between monitor signal and linear prediction signal, as shown in Eq. (9) and Eq. (11). In addition, by using the residual between the original monitoring signal and the linear prediction signal, the residual signal composed of the modulated signal and the noise signal can be effectively obtained, as shown in Eq. (11).

$$x_m(\mathrm{t}) + x_e(\mathrm{t}) = x(\mathrm{t}) + \sum_{k=1}^{p} a(k)x(\mathrm{t}-k) \tag{11}$$

where $x(\mathrm{t})$ is monitor signal; $x_{\mathrm{m}}(\mathrm{t})$ is the modulated signal; $x_{\mathrm{e}}(\mathrm{t})$ is the noise signal; $a(k)$ is the weight coefficient of linear prediction. $a(k)$ can be obtained by linear transformation of AR function of monitoring signal. The process is as follows:

$$r_{xx}(i) = \frac{1}{N}\sum_{n=0}^{N-1} x(n)x(n-1), 0 \le i \le p-1 \tag{12}$$

Therefore, through the linear prediction, the SNR of the modulated signal can be effectively improved, and then, the interference component in the modulated signal can be reduced.

In order to better compare the preprocessing performance of the two algorithms, Gaussian white noise with an SNR of −15 dB is added to the signal model for analysis and verification, as shown in **Figure** 7. The raw signal, TSA signal, AR signal, and their spectrum are shown in **Figure** 7, respectively.

When SNR = −15 dB, the raw signal contains a lot of noise, and the SNR is very low. The spectrum of the raw signal has completely lost the ability to characterize the deterministic signal components. When TSA is used as the preprocessing analysis method, the spectrum of the preprocessed signal contains a certain characteristic frequency within the frequency range of 0–100 Hz, but the spectrum still contains a lot of interference frequencies. At this time, using TSA as the preprocessing method has lost its effect.

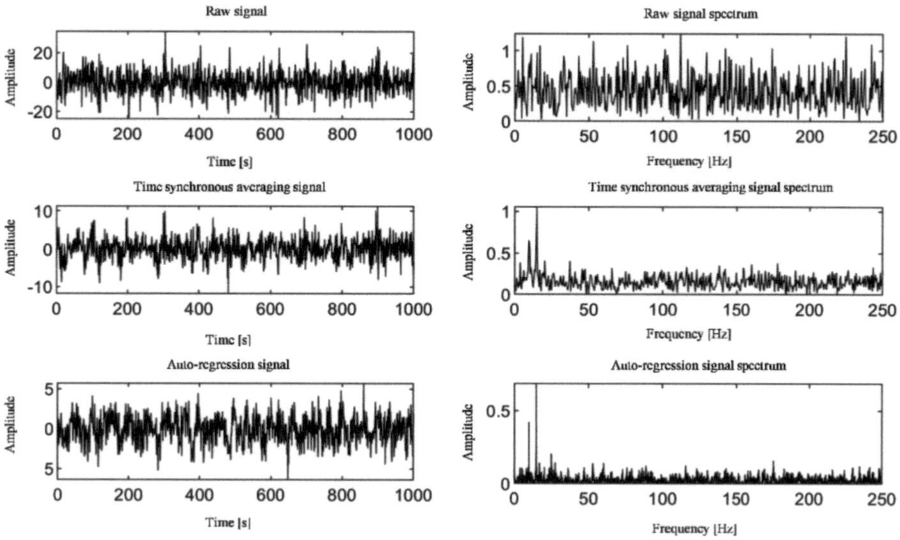

Figure 7.
Comparison of noise reduction. (SNR = −15 dB) [29].

When the linear prediction preprocessing method is adopted, the noise can be effectively reduced according to the waveform of its time-series signal, and the analysis results can be obtained based on the linear prediction as the preprocessing method. There is a low degree of interference in the spectrum, and the characteristic frequency of deterministic signal components can be more accurately located. Therefore, under the condition of low SNR, the preprocessing algorithm based on linear prediction is better than the time-domain average algorithm. In the later sections of this chapter, the preprocessing method based on the AR model was adopted by default.

4. Cases of multi-type typical fault identification fan equipment based on modulation model

4.1 Common failures of multi-type fan equipment

Cyclostationary signal, a kind of widely existing non-periodic and non-stationary signal, is a modulated signal component of rotating machinery. However, the second-order statistical characteristics of cyclostationary signals have certain periodicity, which provides a research basis for cyclic feature extraction of modulated signals. Cyclic feature extraction method is a signal post-processing method based on signal demodulation, which reveals the potential periodicity of the monitoring signal and then obtains more accurate signal modulation information than the traditional signal processing methods. For cyclostationary signals, the enhanced envelope spectrum obtained by Fast-SC [23] could realize cyclic feature extraction. In this section, in order to verify the effectiveness of fault identification and feature extraction methods for different types of fans, experiments of jet fans and axial fans have been carried out. The test rigs of multi-type fan equipment are shown in **Figure 8**. Based on the vibration acceleration signals collected in their experiments, the enhanced envelope spectrum (EES) is calculated, and the characteristic frequency of the fault is extracted.

(a) Jet fan (b) Axial flow fan

Figure 8.
Test rigs of a multi-type fan [1]. (a) Jet fan and (b) axial flow fan.

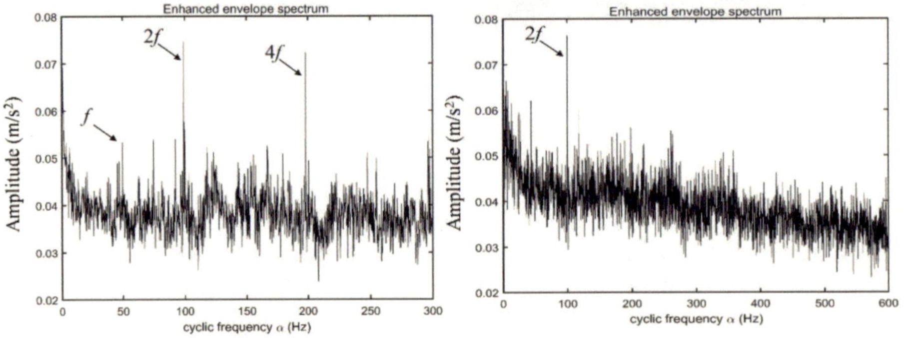

(a) Jet fan (b) Axial flow fan

Figure 9.
Enhanced envelop spectrum in normal working conditions [1]. (a) Jet fan and (b) axial flow fan.

For fault diagnosis, fault characteristics can be used as a criterion for detailed fault diagnosis. Different types of fans have different criteria for fault diagnosis:

For a jet fan, the EES of normal working conditions is shown in **Figure 9a**. It shows that the labeled cyclic frequencies $\alpha = f$, $2f$, $4f$ can represent the characteristic frequencies of the jet fan in normal working conditions, in which f is the shaft frequency of a jet fan. For an axial flow fan, the EES of normal working conditions is shown in **Figure 9b**. It shows that the labeled cyclic frequencies $\alpha = 2f$ are the characteristic frequencies of the axial flow fan in normal working conditions, in which f is the shaft frequency of axial flow fan.

In the fault experiments of a jet fan and an axial fan, bolt looseness faults were set up in the two experimental devices, respectively. The EES of the jet fan in bolt looseness fault is shown in **Figure 10a**. It is indicated that in the labeled cycle frequencies, $\alpha = f$, $2f$, $4f$, $5f$, and $6f$ are the characteristic frequencies of bolt looseness fault of the jet fan. The EES of the axial flow fan in bolt looseness fault is shown in **Figure 10b**. It is shown that the labeled cycle frequencies $\alpha = 2f$, $4f$, $6f$ are the characteristic frequencies of bolt looseness fault of the axial flow fan. The comparison shows that the cyclic feature extraction method can effectively diagnose the bolt looseness fault of an axial flow fan and a jet fan. In addition, the cycle frequency of bolt loosening fault of the jet fan and axial fan is different, so the cyclic feature extraction can be extended to the fault diagnosis of the subway ventilation system.

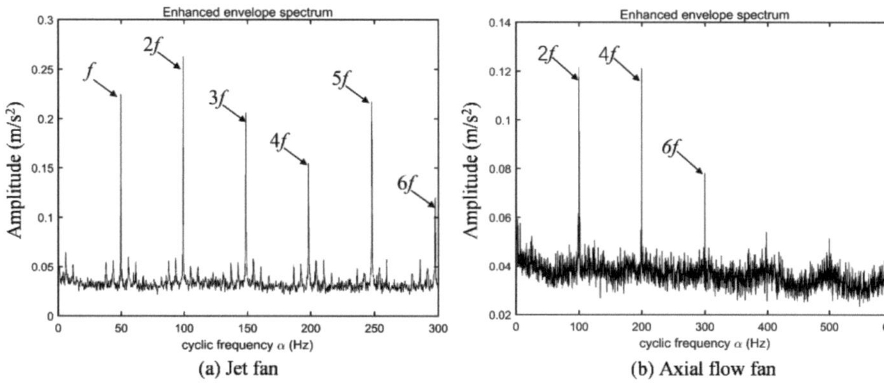

Figure 10.
Enhanced envelop spectrum in bolt looseness fault [1]. (a) Jet fan and (b) axial flow fan.

4.2 Typical faults of the axial flow fan equipment

For different types of typical faults, besides the bolt loosening fault of the axial flow fan device, the EES of blade slight damage fault is also calculated, and the characteristic frequencies of acoustic signals and vibration acceleration signals before and after blades abrasion are extracted, respectively.

For acoustic signals, the EES of normal working conditions is shown in **Figure 11a**. It shows that the labeled cyclic frequencies $\alpha = 2f$, $6f$, $8f$ can represent the characteristic frequencies of acoustic signals in normal working conditions. The EES of damaged blades condition is shown in **Figure 11b**. It is indicated that in the labeled cycle frequencies, $\alpha = 4f$, $6f$, and $8f$ are the characteristic frequencies of blades abrasion fault of the axial fan.

For vibration acceleration signal, the calculated EES before and after blades abrasion are shown in **Figure 12**. It shows that the labeled cyclic frequencies $\alpha = f$, $2f$, $4f$, $6f$, $8f$ can represent the characteristic frequencies of vibration acceleration signals in normal working and blades abrasion conditions. This is because the blades abrasion failure set in the experiment was not destructive damage, so no new characteristic

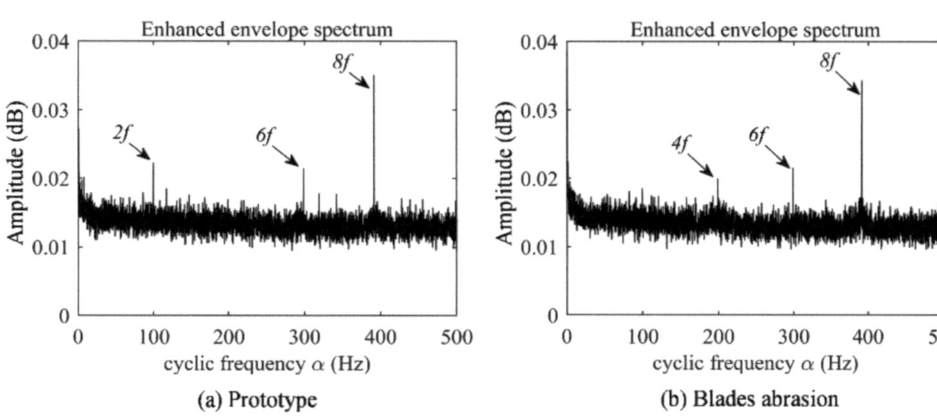

Figure 11.
Enhanced envelop spectrum of acoustic signals. (a) Prototype and (b) blades abrasion.

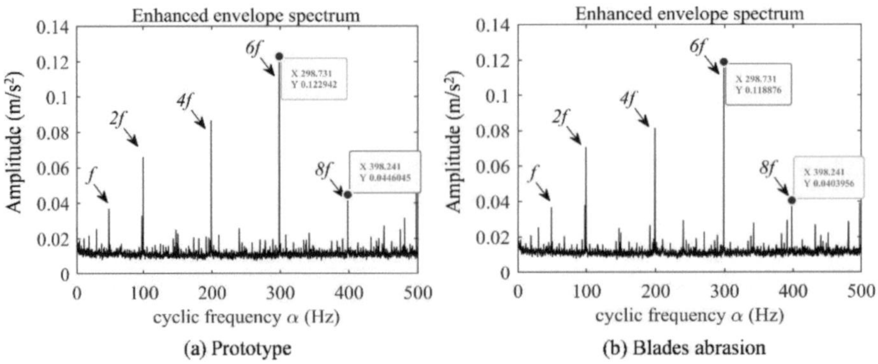

Figure 12.
Enhanced envelop spectrum of vibration acceleration signals. (a) Prototype and (b) blades abrasion.

frequency is generated in the enhanced envelope spectrum under blades abrasion failure. However, compared with normal working conditions, the amplitude of the cyclic frequency caused by blades abrasion fault changes. These amplitude changes can also be used as one of the characteristics of blades abrasion fault, so as to realize the prediction of destructive blades abrasion fault.

5. Conclusion

In this chapter, the key technologies of online monitoring and intelligent diagnosis are discussed. The feature extraction method based on signal demodulation offers a powerful solution to fault identification. In addition, common signal noise reduction methods are researched. Finally, the cases of typical rotating machinery failure were simulated by experiment. The main conclusions are as follows:

1. According to different operational conditions of rotating machinery, its modulation model can be divided into AM signal model and AM-FM signal model. The frequency modulation could be extracted by these models. The establishment of modulation signal model provides the research foundation for a signal demodulation method.

2. Signal preprocessing methods could reduce or eliminate the signal noise, which effectively improves the SNR for further analysis.

3. The fault diagnosis method based on signal demodulation is verified in experiments of bolt looseness and blades abrasion fault. The experimental results show that the amplitudes of cyclic frequency components can reveal the spectral characteristics of the ventilator.

Author details

Yongxing Song
School of Thermal Engineering, Shandong Jianzhu University, Jinan, China

*Address all correspondence to: songyongxing19@sdjzu.edu.cn

References

[1] Song Y, Liu J, Zhang L, Wang Y, Li T, Wu D. The CMM-PWD: A feature extraction method of subway fan under piston wind effect. Revista Internacional de Métodos Numéricos para Cálculo y Diseño en Ingenieria. 2020;**36**:3. DOI: 10.23967/j.rimni.2020.06.004

[2] Liu B. Selection of wavelet packet basis for rotating machinery fault diagnosis. Journal of Sound and Vibration. 2005;**284**:567-582. DOI: 10.1016/j.jsv.2004.06.047

[3] Sur A, Joshi KD, Sharma A, Kaushik TC. Instantaneous frequency extraction of highly nonstationary optical interferometric signal using reassigned smoothed Pseudo Wigner Ville distribution. Optik (Stuttg). 2021; **231**:166436. DOI: 10.1016/j.ijleo.2021.166436

[4] Obuchowski J, Wyłomańska A, Zimroz R. The local maxima method for enhancement of time-frequency map and its application to local damage detection in rotating machines. Mechanical Systems and Signal Processing. 2014;**46**:389-405. DOI: 10.1016/j.ymssp.2014.01.009

[5] Daubechies I, Lu J, Wu HT. Synchrosqueezed wavelet transforms: An empirical mode decomposition-like tool. Applied and Computational Harmonic Analysis. 2011;**30**:243-261. DOI: 10.1016/j.acha.2010.08.002

[6] Liu H, Wang W, Xiang C, Han L, Nie H. A de-noising method using the improved wavelet threshold function based on noise variance estimation. Mechanical Systems and Signal Processing. 2018;**99**:30-46. DOI: 10.1016/j.ymssp.2017.05.034

[7] Williams T, Ribadeneira X, Billington S, Kurfess T. Rolling element bearing diagnostics in run-to-failure lifetime testing. Mechanical Systems and Signal Processing. 2001;**15**:979-993. DOI: 10.1006/mssp.2001.1418

[8] Yang WX, Tse PW. Development of an advanced noise reduction method for vibration analysis based on singular value decomposition. NDTE International. 2003;**36**:419-432. DOI: 10.1016/S0963-8695(03)00044-6

[9] Jiang F, Zhu Z, Li W, Zhou G, Chen G. Fault identification of rotor-bearing system based on ensemble empirical mode decomposition and self-zero space projection analysis. Journal of Sound and Vibration. 2014;**333**: 3321-3331. DOI: 10.1016/j.jsv.2014.03.014

[10] Lei Y, Lin J, He Z, Zuo MJ. A review on empirical mode decomposition in fault diagnosis of rotating machinery. Mechanical Systems and Signal Processing. 2013;**35**:108-126. DOI: 10.1016/j.ymssp.2012.09.015

[11] Li Y, Xu M, Wang R, Huang W. A fault diagnosis scheme for rolling bearing based on local mean decomposition and improved multiscale fuzzy entropy. Journal of Sound and Vibration. 2016;**360**:277-299. DOI: 10.1016/j.jsv.2015.09.016

[12] Žvokelj M, Zupan S, Prebil I. EEMD-based multiscale ICA method for slewing bearing fault detection and diagnosis. Journal of Sound and Vibration. 2016; **370**:394-423. DOI: 10.1016/j.jsv.2016.01.046

[13] Zou H, Xue L. A selective overview of sparse principal component analysis.

Proceedings of the IEEE. 2018;**106**: 1311-1320. DOI: 10.1109/JPROC.2018. 2846588

[14] Fuentes-García M, Maciá-Fernández G, Camacho J. Evaluation of diagnosis methods in PCA-based multivariate statistical process control. Chemometrics and Intelligent Laboratory Systems. 2018;**172**:194-210. DOI: 10.1016/j.chemolab.2017.12.008

[15] Song Y, Liu J, Zhang L, Wu D. Improvement of fast Kurtogram combined with PCA for multiple weak fault features extraction. PRO. 2020;**8**: 1059. DOI: 10.3390/pr8091059

[16] Rahmani M, Atia GK. Coherence pursuit: Fast, simple, and robust principal component analysis. IEEE Transactions on Signal Processing. 2017; **65**:6260-6275. DOI: 10.1109/TSP.2017. 2749215

[17] Li W, Peng M, Liu Y, Jiang N, Wang H, Duan Z. Fault detection, identification and reconstruction of sensors in nuclear power plant with optimized PCA method. Annals of Nuclear Energy. 2018;**113**:105-117. DOI: 10.1016/j.anucene.2017.11.009

[18] Prawin J, Rama Mohan Rao A. An online input force time history reconstruction algorithm using dynamic principal component analysis. Mechanical Systems and Signal Processing. 2018;**99**:516–533. DOI: 10.1016/j.ymssp.2017.06.031

[19] Wang Y, Xiang J, Markert R, Liang M. Spectral kurtosis for fault detection, diagnosis and prognostics of rotating machines: A review with applications. Mechanical Systems and Signal Processing. 2016;**66–67**:679-698. DOI: 10.1016/j.ymssp.2015.04.039

[20] Antoni J, Randall RB. On the use of the cyclic power spectrum in rolling element bearings diagnostics. Journal of Sound and Vibration. 2005;**281**:463-468. DOI: 10.1016/j.jsv.2004.04.007

[21] Antoni J. Cyclostationarity by examples. Mechanical Systems and Signal Processing. 2009;**23**:987-1036. DOI: 10.1016/j.ymssp.2008.10.010

[22] Borghesani P, Pennacchi P, Ricci R, Chatterton S. Testing second order cyclostationarity in the squared envelope spectrum of non-white vibration signals. Mechanical Systems and Signal Processing. 2013;**40**:38-55. DOI: 10.1016/ j.ymssp.2013.05.012

[23] Antoni J, Xin G, Hamzaoui N. Fast computation of the spectral correlation. Mechanical Systems and Signal Processing. 2017;**92**:248-277. DOI: 10.1016/j.ymssp.2017.01.011

[24] Horstmann S, Ramírez D, Schreier PJ. Joint detection of almost-cyclostationary signals and estimation of their cycle period. IEEE Signal Processing Letters. 2018;**25**:1695-1699. DOI: 10.1109/LSP.2018.2871961

[25] Baudin S, Rémond D, Antoni J, Sauvage O. Non-intrusive rattle noise detection in non-stationary conditions by an angle/time cyclostationary approach. Journal of Sound and Vibration. 2016;**366**:501-513. DOI: 10.1016/j.jsv.2015.11.044

[26] Borghesani P, Antoni J. CS2 analysis in presence of non-Gaussian background noise—Effect on traditional estimators and resilience of log-envelope indicators. Mechanical Systems and Signal Processing. 2017;**90**:378-398. DOI: 10.1016/j.ymssp.2016.12.033

[27] Song Y, Liu J, Chu N, Wu P, Wu D. A novel demodulation method for rotating machinery based on time-

frequency analysis and principal component analysis. Journal of Sound and Vibration. 2019;**442**:645-656. DOI: 10.1016/j.jsv.2018.11.024

[28] Song Y, Liu J, Wu D, Zhang L. The MFBD: A novel weak features extraction method for rotating machinery. Journal of Brazilian Society of Mechanical Science Engineering. 2021;**43**:1-16. DOI: 10.1007/s40430-021-03259-z

[29] Song Y. Study on Low Frequency Acoustic Feature Extraction Method of Hydraulic Rotating Machinery Based on Principal Component Analysis. Zhejiang University; 2019